Two-Stroke
Engine Repair
& Maintenance

Two-Stroke Engine Repair & Maintenance

Paul Dempsey

New York   Chicago   San Francisco   Lisbon   London   Madrid
Mexico City   Milan   New Delhi   San Juan   Seoul
Singapore   Sydney   Toronto

5 6 7 8 9 0   DOC/DOC   1 5 4 3 2

ISBN 978-0-07-162539-5
MHID 0-07-162539-9

Sponsoring Editor Judy Bass	**Proofreader** Upendra Prasad, International Typesetting and Composition
Editing Supervisor Stephen M. Smith	**Indexer** WordCo Indexing Services, Inc.
Production Supervisor Pamela A. Pelton	**Art Director, Cover** Jeff Weeks
Project Manager Preeti Longia Sinha, International Typesetting and Composition	**Composition** International Typesetting and Composition
Copy Editor Manish Tiwari, International Typesetting and Composition	

Printed and bound by RR Donnelley.

McGraw-Hill books are available at special quantity discounts to use as premiums and sales promotions, or for use in corporate training programs. To contact a representative, please e-mail us at bulksales@mcgraw-hill.com.

This book is printed on acid-free paper.

About the Author

Paul Dempsey is a master mechanic, and former editor of *World Oil* magazine. He is the author of more than 20 technical books, including *Small Gas Engine Repair*, *How to Repair Briggs & Stratton Engines*, and *Troubleshooting and Repairing Diesel Engines*.

Contents

Introduction

As two-strokes fire every revolution, they are the most powerful engines for their size known. Highly tuned examples develop nearly two hp per cubic inch of displacement and run happily at 11,000-plus rpm. And with only three basic moving parts, two-strokes are the simplest and least expensive form of internal combustion.

Yet, for many owners these little engines are contrivances from hell, cantankerous, difficult to start, and impossible to fix. Drive by a suburban neighborhood on trash collection day and you will find edgers, weed trimmers, and Chinese mini-bikes awaiting pickup at the curbside. The very simplicity of the two-stroke principle makes it unforgiving.

Actually, these engines are easy to live with, if you have the background information and the tools to make a few simple diagnostic tests. And once past the fear of getting their hands dirty, most people find that fixing things is rewarding. Certainly it is more rewarding than spending $85 an hour (the current big-city shop rate) for someone else to do the work. Nor can we continue to discard products that no longer function as they should. That phase of American experience is behind us.

The philosophy of this book was inspired by a lady who visited our class as a substitute teacher lo these many years ago. She was the daughter of a Spanish ambassador to the United States and, during the Second World War, had volunteered to teach Latin American pilots to ferry aircraft across the Atlantic. Although her students were trained pilots, they had not qualified on the large, multi-engined aircraft they would be flying. The students spoke five languages, none of which was English. Flight manuals, written in English, were useless. The lady, whose name I unfortunately cannot recall, realized that her only hope was to simplify instruction. Rather than translate the recipe-book format of manuals, she explained the physics of cockpit instrumentation, how the readings related to the forces acting on aircraft. That sort of knowledge, which cuts to the heart of things, was translatable and memorable. She did not lose a single pilot.

I have tried to do something similar here by stressing how the various components that make up an engine function. Once you understand the basic principles of, say, carburetion, this knowledge becomes a sort of mental tool box that gives you the leverage to repair any carburetor.

The initial chapter describes how two-stroke engines function and the ways these engines have evolved under the pressure of ever-tightening emissions regulations. Encountering a new technology is like meeting someone for the first time. To achieve understanding, to come into sync, requires an appreciation of the forces that have shaped the person.

An internal combustion engine can be thought of as a collection of four systems—ignition, fuel, starting, and those mechanical parts that generate compression. The troubleshooting chapter shows how to isolate a malfunction to a particular system. Fixing on the right system is the basic diagnostic skill that separates mechanics from parts changers. Once you have identified the system at fault, turn to the appropriate chapter for detailed diagnostic procedures and step-by-step repair instructions.

When factory tools are mentioned, they are illustrated so that substitutes can be found or fabricated. And whenever possible, multiple ways of performing the same task are described. Depending on the tools available, you can remove a flywheel by any of three methods. There are at least four ways to separate crankcase castings and several approaches to tuning carburetors.

You will also find much information here on adhesives, sealants, solvents, nylon cord, lubricating oils, and a host of other products that contribute to long-lasting repairs.

This book has more than 100 illustrations, many of them photographs supplied by my good friend, Robert Shelby. STIHL, Tanaka, Dorman, and several other manufacturers were kind enough to allow illustrations from their parts and shop manuals to be used.

Paul Dempsey

Two-Stroke Engine Repair & Maintenance

1

Fundamentals

There comes a time when work stops and the mechanic becomes abstracted, distant from the task at hand. Something about the machine does not conform to the picture in the mechanic's mind. Images flash by until he or she finds one that most closely conforms to actual conditions. Once that is done, repairs can begin.

Constructing visual images is what mechanics do; the other stuff is mere wrench-twisting.

This chapter provides grist for these mental images. Because the material must be conveyed in words, it tends to be abstract. But once you can picture how these engines work, you will have made the first step in the journey to becoming a real mechanic.

Spark-ignition engines operate in a cycle consisting of four events: intake, compression, expansion (or power), and exhaust. A fresh charge of air and fuel is inducted into the cylinder, which then is compressed by the piston and ignited by the spark plug. The pressure created by combustion reacts against the piston to generate torque on the crankshaft. The spent gases then exhaust into the atmosphere.

Four-stroke-cycle engines require four up and down strokes of the piston, or two full crankshaft revolutions, to complete the cycle. Two-stroke-cycle engines telescope events into two strokes or one crankshaft revolution. For convenience we abbreviate the terms to four-cycle or four-stroke, and two-cycle or two-stroke.

Two-cycle operation

Focus on the piston. The double-acting piston works in both directions to compress the air-fuel mixture in the cylinder above it and in the crankcase below it. The piston and connecting rod convert a portion of the heat and

1

energy released by combustion into mechanical motion that turns the crank-shaft. Were that not enough, the piston also acts as a slide valve to open and close exhaust, transfer and (in some applications) intake ports. Because it works so hard, the piston is the first mechanical part to fail on two-cycle engines.

Third-port engines

Third- or piston-ported engines have three ports cast or milled into their cylinder liners. The inlet port admits fuel to the crankcase, the transfer port conveys fuel from the crankcase into the combustion chamber, and the exhaust port opens to the atmosphere.

First, let's look at events above the piston during a full turn of the crank-shaft. In Fig. 1-1A the piston approaches the upper limit of travel, or top dead center (TDC), and has compressed the air-fuel mixture above it. The piston has also uncovered the inlet port to admit fuel and air from the car-buretor to the crankcase. Figure 1-1B illustrates the beginning of the power stroke under the impetus of expanding combustion gases. As the piston falls, it first uncovers the exhaust port (Fig. 1-1C) and, a few degrees of crankshaft rotation later, the transfer port (Fig.1-1D). Fuel and air pass through the transfer port and into the cylinder bore.

Meanwhile, much is happening in the crankcase. As the piston falls on the power stroke, it partially fills the crankcase, reducing its volume, as shown in Fig. 1-1C. Since the piston now covers the inlet port, the pressure of the air-fuel mixture trapped in the case rises.

Near bottom dead center (BDC) the piston uncovers the transfer port and the pressurized fuel mixture passes through this port to the upper cylinder (Fig. 1-1D). The piston then rounds BDC and begins to climb, an action that simultaneously compresses the mixture above the piston and creates a par-tial vacuum under it. Once the inlet port opens, atmospheric pressure forces fuel and air from the carburetor into the crankcase.

A problem with third-port engines is fuel reversion. At low speeds the crankcase fills to overflowing. When the piston reverses at the top of the stroke, some of the charge can flow back through the inlet port to the car-buretor. A fog of oily fuel hovers around the air cleaner, dirtying the engine and playing havoc with carburetor metering.

Reed-valve engines

Although third-port engines are still encountered, many manufacturers prefer to control crankcase filling with a reed valve installed between the carburetor and crankcase. The valve, similar to the reed on musical instruments, opens and closes in response to crankcase pressure (Fig. 1-2). Utility engines make do with a single reed, or pedal, athwart the intake port (Fig. 1-3). High-performance

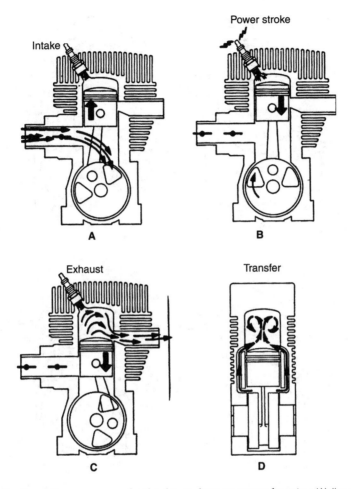

FIG. 1-1. *Operating sequence of a third-port, loop-scavenged engine.* Walbro

engines employ a tent-like valve block with multiple reeds. This arrangement provides a large valve area for better crankcase filling (Fig. 1-4).

For mini-two-strokes, the position of the carburetor indicates the type of inlet valve: when a reed is present, the carburetor mounts on the crankcase (Fig. 1-5). Third-port engines mount their carburetors on the cylinder barrel in line with the inlet port, as shown in Fig. 1-6. Being able to recognize the presence of a reed valve without disassembling the engine is useful, since the reed can malfunction. Should the pedal split or fail to seal, the engine will not start.

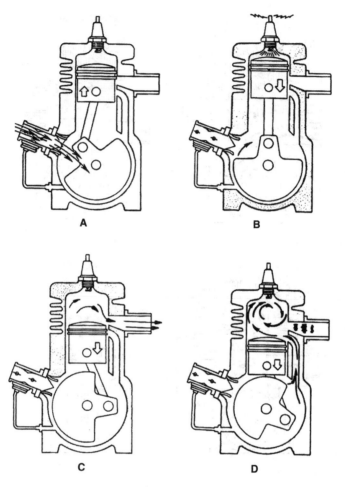

FIG. 1-2. *Operating sequence of a reed-valve engine that in the example shown employs loop scavenging. The small tube on the lower left of the drawing transfers crankcase pressure pulses to the fuel pump.* Deere and Company

But the rule about the carburetor location does not necessarily apply to larger engines. Some European motorcycles had crankcase-mounted carburetors that fed through a rotary valve in the form of a partially cutaway disk, keyed to crankshaft. Model airplane engines and a few vintage outboards use a slotted crankshaft to the same effect.

Motorcycle engines often combine a third port with an integral reed valve. The port controls timing and the reed prevents backflow through the carburetor. Although the reeds impose a pressure drop, midrange torque benefits.

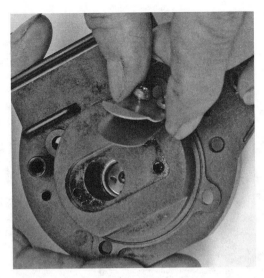

FIG. 1-3. *Reed valves for handheld engines generally have a single pedal backed by a guard plate to limit deflection.* Robert Shelby

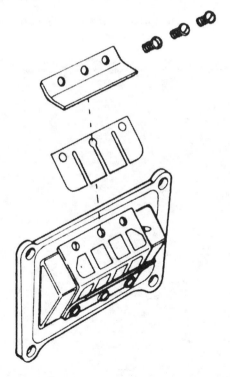

FIG. 1-4. *Multiple pedals are standard on high-performance engines. While there has been considerable experimentation with fiberglass, carbon fiber, and other high-tech materials, spring-steel pedals appear to work as well as any.* Tecumseh Products Co.

FIG. 1-5. *Reed-valve engines mount their carburetors low on the crankcase.* Robert Shelby

FIG. 1-6. *Carburetors for third-port engines attach to the cylinder. Some of these engines incorporate a reed valve in the third port.* Robert Shelby

Scavenging

Scavenging is the term for purging the cylinder of exhaust gases. Unlike a four-cycle engine, which devotes a full stroke of the piston to clear the cylinder, a two-stroke must scavenge during the 100° or so of crankshaft rotation that the exhaust port remains open.

Blowdown As the piston falls, it first uncovers the exhaust port and then, 5° or 10° of crankshaft rotation later, the transfer port. Blowdown occurs during this brief period that, at wide-open throttle, occupies no more than one or two thousandths of a second. In spite of its brevity, the blowdown phase is the primary mechanism for evacuating the cylinder.

The rapid opening of a port releases a high-pressure slug of exhaust gas that trails a low-pressure zone or wave in its wake. Cylinder pressure momentarily drops below atmospheric pressure. Responding to the pressure differential, the fresh charge moves through the transfer port to fill the cylinder. At part throttle, crankcase pressure is less than cylinder pressure. Were it not for the drop in cylinder pressure that accompanies blowdown, two-cycle engines would not run.

The need to accelerate exhaust gases quickly explains why exhaust ports for high-performance engines are rectangular rather than round. It also explains why we must keep these ports and mufflers free of carbon accumulations.

Exhaust tuning When a high-pressure wave encounters a solid obstacle or an abrupt change in direction in the exhaust plumbing, it rebounds back to the exhaust port. These waves oscillate at the speed of sound and at a frequency determined by engine rpm. Where space permits, the length of the exhaust system can be tuned to reflect a high-pressure wave back to the exhaust port just as the cylinder fills to overflowing. The wave rams any fuel that spills out of the port back into the cylinder where it belongs. Of course, this works only over a narrow rpm range; at other speeds the wave can arrive early to the detriment of cylinder filling. In a similar manner, the intake tract can be tuned to maximize crankcase filling.

Spatial constraints make tuned exhaust and intake systems impractical for handheld equipment. About all that can be done is to arrange for a small boost from third- or fourth-order wave harmonics.

Charge scavenging What exhaust gas remains in the cylinder after blowdown must be scavenged by the fuel charge, which enters the cylinder at velocities as high as 65 m/s. Charge scavenging takes two forms, neither of which can entirely eliminate short-circuiting.

Short-circuiting Short-circuiting is the term for the way incoming fuel escapes out the exhaust, as if one were trying to fill a leaky bucket. Most of the leak can be laid to symmetrical timing.

Because piston motion controls port timing, the timing is symmetrical around BDC. For example, an exhaust port that opens 60 crankshaft degrees before bottom dead center must remain open for 60° after BDC. The exhaust port opens a few degrees before the transfer port. Otherwise, the cylinder would not blow down and very little fuel would be delivered. But opening the exhaust port early means that it stays open throughout the entire fuel transfer process.

The open exhaust port acts as an escape hatch for incoming fuel. How much fuel escapes combustion varies with port geometry, rpm, and throttle position. An idling motorcycle short-circuits as much as 70% of its fuel out the exhaust. On average, two-stroke engines waste between 25% and 35% of their fuel in this manner.

Cross scavenging Readers with long memories may recall the deflector pistons that were once standard ware on these engines (Fig. 1-7).

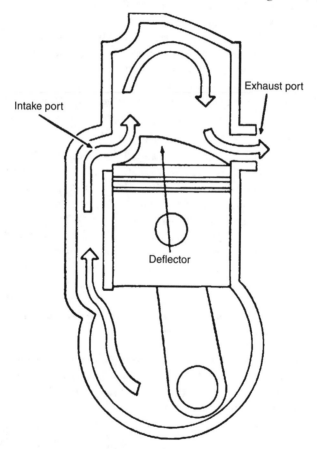

FIG. 1-7. *Cross-scavenged engines have the intake port 180° opposite the exhaust port. Incoming gases rebound upward off the deflector on the piston crown to give some protection against short-circuiting.*

The fuel charge enters through a single transfer port, rebounds upward off the deflector, and drives residual exhaust gases out the exhaust port. While this design works well at moderate speeds, at high speeds, the deflector can run hot enough to ignite the mixture. Nor does the simple trajectory made by the incoming charge impact the area just above the exhaust port, which remains a haven for exhaust gas. Other factors that mitigate against cross scavenging include the awkward shape of the combustion chamber and the weight penalty imposed by the deflector. But the single transfer port simplifies foundry work, which explains why American outboard manufacturers were among the last to abandon this approach.

Loop scavenging Current practice, based on work carried on in Germany during the 1920s, is to use loop, or Schn_rle, scavenging. Multiple transfer ports are arranged around the cylinder periphery with their exit ramps angled to impart swirl to the charge (Figs. 1-8 and 1-9). The miniature cyclone fills the whole combustion chamber, sweeping exhaust gases out ahead of it. In addition, the rapidly spinning mass of fuel and air has integrity, that is, it hangs together so that less fuel short-circuits.

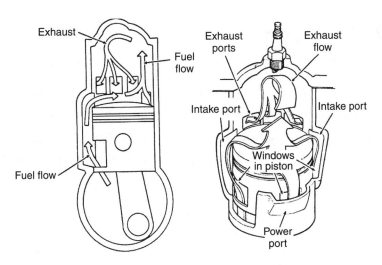

FIG. 1-8. *In a loop-scavenged engine the fuel charge enters through multiple transfer ports (called intake ports here) arranged around the periphery of the cylinder. Port exit angles to give swirl to the charge, which reduces short-circuiting. Windows in the piston skirt are an optional feature.* OMC

FIG. 1-9. *This cylinder has what are sometimes called finger ports. That is, the transfer ports are open to the bore along their whole length. Looking carefully one can see the angled exit ramp at the upper end of lower port.* Robert Shelby

Displacement

We class ships by tonnage, houses by square footage, and engines by the volume the piston displaces as it moves between centers. All things equal, an engine should develop power in proportion to its displacement.

Displacement = bore × bore × number of cylinders × stroke × 0.7858

For example, the Tanaka series TBC-2501 has a 34-mm bore and a 27-mm stroke. To perform the calculation, square the bore and multiply by the stroke:

$$34 \times 34 \times 1 \times 27 \times 0.7858 = 24526.39 \text{ mm}^2$$

To convert to cubic centimeters, divide by 1000:

$$24526/1000 = 24.5 \text{ cc}$$

To convert to cubic inches, multiply cubic centimeters by 0.061, which in this example gives 1.50 CID (cubic-inch displacement). To work the conversion the other way, multiply the CID by 16.387 to arrive at cubic centimeters.

Compression ratio

The compression ratio (CR) describes the amount of "squeeze" the piston applies to the air-fuel mixture prior to combustion. It is arrived at by dividing total cylinder volume, that is, the volume with piston at BDC, by the volume that remains when the piston rises to TDC. The latter figure is the clearance volume. Normally we take the manufacturer's word for CR, since determining the clearance volume can become a bit hairy, especially when a domed piston is fitted.

Some manufacturers express CR as just described, which is geometrically accurate and yields impressively high numbers. Others provide the effective ratio, calculating swept volume as the volume the piston transverses after the exhaust port closes. Obviously, there can be no compression with an open exhaust port. Effective CRs range from 6 to more than 8:1.

Up to a point, the higher the compression ratio the better. The limit is imposed by the tendency of the fuel, a tendency made worse by the presence of oil, to detonate. Normal combustion is an orderly process, initiated by the spark and moving out to fill the combustion chamber. Detonation occurs when the tag ends of the fuel charge, compressed and heated by the expanding flame front, suddenly explode (Fig. 1-10). Cylinder pressures skyrocket and, if detonation persists, the piston melts and crankpin bearings hammer flat.

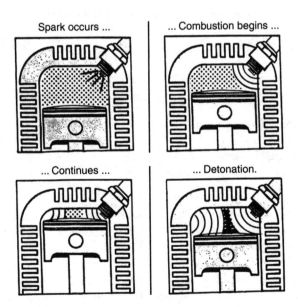

FIG. 1-10. *As shown in this Champion Spark Plug drawing, detonation is a maverick form of combustion initiated late in the process, after normal ignition.*

In addition to cylinder compression, two-strokes also develop crankcase compression. Since the work of compressing the air/fuel charge prior to delivery absorbs energy that could be better used to turn the crankshaft, designers limit crankcase compression to between 1.3 and 1.6:1. Pressures rarely exceed 6 psi.

Torque and horsepower

Near the end of the eighteenth century, James Watt observed that a mine pony tethered to a turnstile could lift 550 lb one foot per second, or 33,000 lb per minute. Horsepower was a brilliant sales tool that put steam engines into a context that potential customers could understand. In metric notation, one horsepower equals 0.746 kilowatt (kW).

Torque, or the instantaneous twisting force on the crankshaft, is the active component of horsepower. One foot-pound of torque is a twisting force generated by a one-pound weight on the end of a bar one foot long. Expressed metrically, 1 ft/lb equals 1.36 Newton meters (Nm) or 0.14 kilogram meters (kgm).

To determine torque output, researchers mount the engine on a dynamometer and measure the braking force required to bring the engine to a halt. Once torque is known, the conversion to brake horsepower (bhp) is simple:

$$bhp = (torque \times rpm \times 2\ \pi)/33{,}000$$

Thus, an engine that produces 2 ft/lb of torque at 7000 rpm has a power output of

$$bhp = (2 \times 7000 \times 2 \times 3.14)/33000$$

$$bhp = 2.66$$

We feel the effects of torque as the ability of vehicles to accelerate and as the refusal of handheld tools to bog down under sudden loads. Internal combustion engines generally develop their maximum torque at about two-thirds throttle.

$$hp = (torque \times rpm)/5253$$

Peak hp occurs at full throttle or at some close approximation to it. The limiting factor is friction, especially friction between the piston rings and cylinder walls, which increases with speed even more rapidly than horsepower.

The horsepower rating of engines remains a formidable sales tool that, in the absence of standards, can be easily manipulated. Reputable small-engine manufacturers arrive at their horsepower numbers in accordance with the Society of Automotive Engineers (SAE) protocol J-1349. The test engine is tuned to laboratory precision, a process that includes decarbonization after break-in, and its output measured at an ambient temperature 77°F (25°C) and an elevation of 100 m (328 ft) above sea level.

Since most engine makers follow this protocol, advertised horsepower can be used as a means of comparison between models and brands. But laboratory levels of tuning boost horsepower beyond levels experienced in the field. As a practical matter, customers can expect no more than about 85% of the horses promised.

Premix

Most modern two-strokes require 89 octane gasolines to prevent detonation. And the fuel should not be more than a few weeks old. As gasoline ages, the light hydrocarbons evaporate, leaving varnish and gums behind. A fuel stabilizer, such as Briggs & Stratton Fresh Start, will preserve gasoline for as long as 24 months. However, stabilizers are not retroactive—stale fuel cannot be restored to its original state. Nor should gasolines with high alcohol (methanol or ethanol) content be used. Alcohol collects water that can promote corrosion and tends to cause oil to drop out of the premix.

When water is a problem, the fuel can be strained through a chamois or a water-separator funnel available from marine-supply houses.

The safest choice is to follow manufacturer's recommendations for lube oil. Approved lubricants for air-cooled two-strokes conform to JASC M345/FD and ISO-LEGD requirements. Some of these oils also carry Rotax snowmobile Test 2 certification.

It's interesting to note that Rotax warns against using pure synthetic oils in engines that stand idle for more than a few days at a time. The oil film left by synthetics drains off and exposes parts to corrosion. Some, but not all, manufacturers share the same concern. In-house lubricants from Husqvarna and Echo are blends of synthetic and traditional mineral oils.

The fuel-oil mixture as specified by the engine manufacturer (and not the oil refiner) should be used. Too little oil makes quick work of the engine; too much reduces the octane of the premix, scores pistons with carbon flakes and overheats catalytic converters. Most handheld equipment require a 50:1 mix, although some need more oil at 32:1. Vintage mosquito foggers, with their plain-bearing connecting rods, run at ratios of 16 or even 12:1.

Accurate measurement calls for a graduated flask and care to sight over the top of the oil and not the meniscus that clings to the sides of the flask. Measure the amount of gasoline at the pump, pour the oil into the container, and shake vigorously. Since oil settles out over time, give the container a few shakes before refueling (Table 1-1).

Table 1-1
Gasoline-to-Oil Ratios

Ratio	U.S.		Metric	
	Gasoline (gal)	Oil (oz)	Gasoline (l)	Oil (ml)
24:1	1	5.3	4	167
	2	11.0	8	333
32:1	1	4.0	4	125
	2	8.0	8	250
40:1	1	3.2	4	100
	2	6.4	8	200
50:1	1	2.5	4	80
	2	5.0	8	160
100:1	1	1.3	4	40
	2	2.6	8	80

Cooling

Only about 20% of the heat released during combustion is transformed into mechanical motion. The remainder goes out to the atmosphere by way of the exhaust and cooling system.

Conduction

Heat generated by combustion and friction produces molecular vibration in the surrounding metal, which passes from one molecule to the next in a process known as conduction. Since both thermal and electrical conduction depend upon the same mobile electrons, good electrical conductors, such as copper and aluminum, also make good thermal conductors.

The rate of heat transfer P depends upon

- The difference in temperature ($T2$ hot − $T1$ cold), also known as the thermal gradient,
- Thermal conductivity k of the material,
- Area A of the material,
- The length L of the thermal path.

$$P = kA(T2 - T1)/L$$

To cool a cylinder, we want to use a material with good thermal conductivity (k), add fins to increase its area (A), and make the cylinder walls as thin as possible to reduce the length (L) of the thermal path. We can't do much about the thermal gradient, but it is already quite steep, since combustion temperature is more than an order of magnitude greater than the temperature of the surrounding air.

As a point of interest, the LeGnome rotary aircraft engine, widely used during the First World War, holds what must be a Guinness record for short thermal paths. Its steel cylinders were only a 0.5-mm (0.13-in.) thick. The fins provided the hoop strength necessary to withstand combustion forces. In addition, each piston was fitted with a brass ring that flexed to accommodate cylinder distortion. It's not surprising that these engines required overhaul after as little as 10 hours of flying time.

Once heat has passed through the cylinder head and barrel, it must be dissipated into the atmosphere.

Radiation

Thermal radiation is an electromagnetic wave phenomenon that, in several respects, mimics the behavior of light. Infrared waves emanate at right angles from a warm surface. As temperature increases, the dominate wave frequency shifts to the visible spectrum. The progression goes from cherry red to white heat.

Radiation requires a target, preferably a dark object, to convert itself into heat. A wood stove warms one standing before it, while the surrounding air remains cold. And, as Russian peasants have known since time immemorial, a dark winter coat absorbs more heat than a light-colored one.

Table 1-2 lists emissivity coefficients for various states of aluminum at room temperature relative to the perfect emissor, or "black body." Such an object would have an emissivity coefficient of 1.

From this table, you can see that the rough and as-cast finishes on aluminum castings are pretty well ideal for heat transfer by radiation.

Convection

Convection, or the movement of thermal energy in fluids, can be demonstrated by the behavior of water heated in a pan. The water near the bottom

Table 1-2
Emissivity Coefficients

Material	Emissivity Coefficient (1.0 = the Highest Value)
Aluminum	
anodized	0.77
as cast	0.93
oxidized	0.20–0.31
polished	0.04–0.06
Steel	
polished	0.19
rusted	0.07

of the pan rises and is replaced by cooler and heavier water. The same phenomenon occurs between adjacent cylinder fins: heated air moves away from the fins to be replaced by cooler air.

The cooling effect is intensified if we pressurize air with a fan and direct it over the fins by means of shrouding. Handheld and stationary engines employ this form of cooling, known as forced-air convection. Energizing air with a fan means that cylinder fins can be closely spaced and as thin as casting techniques permit. The fan forces air through the narrow interstices.

Motorcycles depend upon their forward movement to create a cooling draught. Free-air cooling eliminates the power drain of a fan, but works only if the machine is moving and moving without the assistance of a tailwind. Because of the relatively low air velocity, fins are widely spaced and, in partial compensation for the loss of area, tend to be tall (Fig. 1-11).

Radiation plays an important role in free-air cooling. When the vehicle is moving, radiation accounts for between a tenth or a sixth of the heat transfer. But if the bike is stationary, temperatures rapidly increase. Once past a certain threshold, the heat energy released by radiation goes up as the fourth power of the absolute, or Kelvin, temperature. K = 273.15° + °C. If cylinder-head temperature doubles from 300° to 600° (the equivalent of 328°), the rate of heat rejection is multiplied 16 times. It is no exaggeration to say that radiation makes free-air cooling practical.

Almost any engine can be made to run cooler. Cylinder fins should be clean and free of oil, which acts as a thermal insulator. Shrouds should fit tightly over the fins without air leaks. Unshrouded aluminum castings radiate better with a *light* coat of paint. Some professionals use Rustoleum black enamel, cut with gasoline to eliminate the gloss. But color has less of an effect than the type of paint used. Light colors are used on the interior surfaces of microwave ovens. Epoxy and urethane paints deliver 90% or more of black body radiation at room temperatures; enamels do less well at around 83%.

Emissions

On the face of it, concern about emissions from thimble-sized engines seems frivolous. But those who make their livings with handheld tools might see things differently. Or at least, they should. According to the California Air Resources Board (CARB), an hour of weed whacking is the emissions equivalent of driving 834 miles in a modern automobile. It's also true that two-cycle exhaust has a serious impact upon heath in the developing world. An estimated 70 to 100 million motorcycles, mopeds, tuk-tuks, and tricycles vie for space on the streets of Asiatic cities.* These engines are believed to account for

---

*B. Wilson, "Direct Injection as a Retrofit Strategy for Reducing Emissions from Two-Stroke Cycle Engines in Asia," Dec., 2002, SAE 80523-1374.

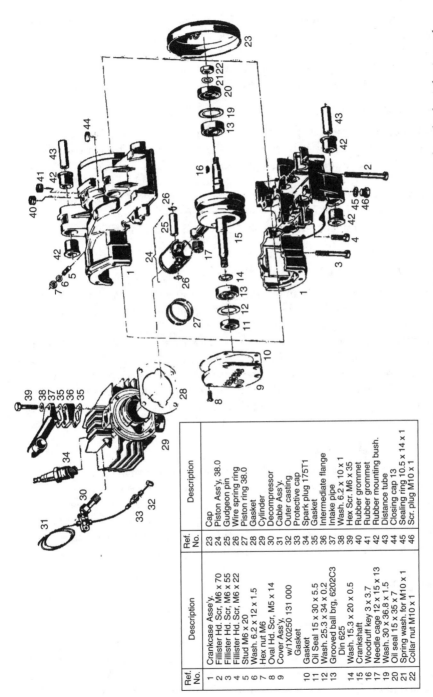

Ref. No.	Description	Ref. No.	Description
1	Crankcase Ass'y.	23	Cap
2	Fillister Hd. Scr, M6 x 70	24	Piston Ass'y, 38.0
3	Fillister Hd. Scr, M6 x 55	25	Gudgeon pin
4	Fillister Hd. Scr, M6 x 22	26	Wire spring ring
5	Stud M6 x 20	27	Piston ring 38.0
6	Wash. 6.2 x 12 x 1.5	28	Gasket
7	Hex nut M6	29	Cylinder
8	Oval Hd. Scr, M5 x 14	30	Decompressor
9	Cover Ass'y,	31	Cable Ass'y.
	w/1X0250 131 000	32	Outer casting
	Gasket	33	Protective cap
10	Gasket	34	Spark plug 175T1
11	Oil Seal 15 x 30 x 5.5	35	Gasket
12	Wash. 25.3 x 34 x 0.2	36	Intermediate flange
13	Grooved ball brg, 6202C3	37	Intake pipe
	Din 625	38	Wash. 6.2 x 10 x 1
14	Wash. 15.3 x 20 x 0.5	39	Hex Scr. M6 x 35
15	Crankshaft	40	Rubber grommet
16	Woodruff key 3 x 3.7	41	Rubber grommet
17	Needle cage 12 x 15 x 13	42	Rubber mounting bush.
19	Wash. 30 x 36.8 x 1.5	43	Distance tube
20	Oil seal 15 x 35 x7	44	Closing cap 13
21	Spring wash. for M10 x 1	45	Sealing ring 10.5 x 14 x 1
22	Collar nut M10 x 1	46	Scr. plug M10 x 1

FIG. 1-11. *The Ficket & Sachs 505 depends upon the forward motion of the moped for cooling. Because the cylinder is horizontal, the fins run longitudinally along the length of the barrel.*

17

70% of India's hydrocarbon emissions, 46% of the nation's carbon monoxide emissions, and a large fraction of its particulate matter.

The emissions data presented in the following discussion are based on a European Commission study of three two-stroke engines (one trimmer and two chainsaws) and two four-stroke trimmers.

Hydrocarbons

Short-circuiting helps explain why the two-stroke engines produced as much as four times more hydrocarbons (HC) than their four-stroke counterparts. Rich mixtures exacerbate the problem. Other sources of HC include quenching, as the flame front cools in contact with metal surfaces. The worst offender in this regard is the narrow crevice between the piston and the cylinder bore in the area above the upper ring.

The primary effect of HC is smog, although several mutagenic and carcinogenic compounds are present in the mix.

Carbon monoxide and carbon dioxide

Carbon monoxide (CO) is an odorless and colorless gas that, when inhaled, displaces oxygen in the blood stream. Low exposures result in nausea and headaches; higher levels of exposure are lethal.

The two-strokes in the European tests produced about the same level of CO as the four-strokes. Carbon monoxide results from incomplete combustion. When sufficient oxygen is present, carbon combines with a second oxygen molecule to produce carbon dioxide (CO_2), the infamous global warming gas. Other studies have demonstrated a powerful linkage between CO and carburetor settings. Rich mixtures can emit an order of magnitude more CO than extremely lean mixtures.

Oxides of nitrogen

Oxides of nitrogen (NOx) (rhymes with "socks") is a blanket term for the various oxides of nitrogen that react with the environment to cause smog, ground-level ozone, and acid rain. Fortunately, two-cycle engines do not attain the high combustion temperatures required for large-scale NOx formation. In the European Commission tests, two-strokes produced about half the oxides of nitrogen as the four-strokes.

Particulate matter

Particulate matter (PM) is the nongaseous component of exhaust, consisting of tiny spherules of carbon, loosely bound in clusters or chains. Spewed out the exhaust like shotgun pellets, these solids imbed themselves deep into lung tissue where they pose a serious heath hazard. In addition, PM particles often coalesce around a powerful carcinogen known as the soluble organic fraction (SOF) of polyaromatic hydrocarbons. According to the

American Lung Association, PM inhalation from all sources accounts for 50,000 deaths annually in the United States.

The European study did not measure particulate matter, since it is not currently regulated. However, Finnish researchers found that the volume of PM emitted by a 46-cc, two-stroke chainsaw "was relatively large compared to particulate emissions from automobiles."[*] More critical was their finding that 95% of these solids fell into the SOF category. Overoiling should be avoided: a 6% oil-gasoline mix generates three times more PM than a 2% mixture.

Regulations

As far as the North America is concerned, the impact of emissions standards on two-cycle engines is pretty well confined to garden equipment and other nonroad applications. Almost no road-legal two-stroke motorcycles survive in this market.

In 1995 CARB cut emissions levels by 30% for nonroad small engines sold in California under what came to be known as Tier 1 regulations. Several northeastern states followed California's lead. Only three pollutants were of concern— CO, NOx, and HC—with the latter two combined into a gram per kilowatt hour (1 kW/h = 1.32 hp) limit.

Two years later, the Federal Environmental Protection Agency (EPA) followed up with its own Phase 1 rules that reduced emissions for nonhandheld engines sold elsewhere in the country by 32%.

Although several manufacturers objected, Tier 1/Phase 1 regulations caused no great hardship. Carburetors were calibrated leaner and restrictor caps were fitted to the mixture-adjustment screws to prevent tampering.

CARB then announced that it would reduce small-engine emissions by another 80%. Tier 2 was to come into effect in 1999. The EPA proposed similar limits for Phase 2, but softened the impact by spreading compliance over a 3-year period beginning in 2002. Both agencies permitted manufacturers to offset noncompliant engines with cleaner engines, bank emissions credits, and trade credits with competitors. Tier 2/Phase 2 also includes a provision for a second emissions test to assure compliance as equipment ages. The "Emission Compliance Period" equates with the useful life of the product and ranges from 50 to 500 hours.

The Portable Power Equipment Manufacturers Association (PPEMA) saw these regulations as an attack on their base. An 80% reduction in emissions seemed unrealistic, especially in view of the profit squeeze imposed by discount houses. Lowe's, Home Depot, Walmart and the like account for nearly three-quarters of garden equipment sales. An Echo manufacturing executive would later confide to *Fortune* magazine that the struggle cost him his hair and 35 lb of weight gain.

---

[*]S. Laanti, J. Sorvari, E. Elonen, M. Pitkänen, "Mutagenicity and PAHS of Particulate Emissions of Two-Stroke Chainsaw Engines," Jan., 2001, SAE 1825-4246.

The trade group rolled up its collective sleeves and went to work on the political front. The initial target was CARB. PPEMA asked for a 50% cap on emissions, a figure that could be achieved with conventional technology. But Tanaka and RedMax had another agenda. Thanks to their recently developed stratified-charge engines, these companies were confident that they could meet the 80% cap. During a 10-hour meeting they urged CARB to hold firm. And hold firm CARB did, but as a sop to PPEMA, delayed implementation of Tier 2 for a year.

Some two-stroke makers failed to meet these standards and could no longer remain in business. McCulloch, once the king of chainsaws, filed for bankruptcy in 1999 and is now in the hands of a Taiwanese company.

Efforts to mollify or delay EPA Phase 2 regulations met the same fate. According to industry insiders, the battle was lost, in great part, because of the deflection of John Deere and Co. Initially Deere had opposed Phase 2, but then lobbied to keep the regulations intact. Deere had its own version of stratified scavenging. In disarray, PPEMA closed its doors in 2001. Most of its members migrated to the Outdoor Power Equipment Institute.

EPA Phase 3, scheduled to take effect in the 2012 model year, will further tighten emissions for lawnmowers and portable generators. And following CARB's lead, the EPA will impose evaporative limits, which require non-permeable fuel tanks and lines.

The European Union has signaled its intent to regulate CO_2 emissions from all sources, including small engines. As things stand today, the only way to limit carbon dioxide is to reduce fuel consumption.

Ever-tightening regulatory constraints leave manufacturers with two options: switch to four-cycle engines or else figure out ways to clean up the traditional product (Table 1-3).

The four-cycle option

Ryobi invested $10 million in the four-cycle engine that powered the first string trimmer certified under CARB Tier 2. The company subsequently collaborated with RedMax-Komatsu Zenoah to produce a heavy-duty version of the engine. Honda, recognizing the opportunity, entered the market with its overhead cam GX22 and GX31. Briggs & Stratton followed with its 34-cc Fource engine. Thanks to mist-type oiling systems and diaphragm carburetors, these little four-bangers operate at any angle.

But four-cycle micro engines have their detractors. According to a spokesman for Robin America, a four-stroke 25-cc engine is, on average, 1.1 lb heavier than the equivalent two-stroke. And that's a conservative estimate: the 29-cc Craftsman model 79197 four-stroke trimmer tilts the scales at 20 lb,

Table 1-3
U.S. Emissions Standards for Small Engines in Handheld and Nonhandheld Applications

CARB Tier 2

	2000	2001	2002	2003	2004	2005	EPA Phase 2	
							2006	2007 and forward
0–65 cc	HC + NOx = 729 CO = 536	↓	↓	↓	↓			
< 50 cc						HC + NOx = 50 CO = 536 PM = 2.0		↓
50–80 cc						HC + NOx = 72 CO = 536 PM = 2.0	↓	↓

EPA Phase 1

	2002	2003	2004	2005	2006	2007 and forward
			EPA Phase 2			
Handheld	HC + NOx = 238 CO = 805	HC + NOx = 175 CO = 805	HC + NOx = 113 CO = 805	HC + NOx = 113 CO = 805	HC + NOx = 50 CO = 805	HC + NOx = 50 CO = 805

Table 1-3
U.S. Emissions Standards for Small Engines in Handheld and Nonhandheld Applications (*Continued*)

	EPA Phase 1	EPA Phase 2			
Handheld 20–50 cc	↓	HC + NOx = 143 CO = 603 ↓	HC + NOx = 119 CO = 603 ↓	HC + NOx = 96 CO = 603 ↓	HC + NOx = 72 CO = 603 ↓
Non-handheld < 66 cc	HC + NOx = 50 CO = 610	↓			

EPA Proposed Phase 3 to go into effect between 2010 and 2014.

22

less fuel and oil. In contrast, a Husqvarna two-stroke trimmer of the same displacement weights 10.6 lb. The weight penalty comes into sharper focus when we realize that these little four-strokes develop 30 to 50% less power than equivalent two-strokes.

The valve gear required by four-stroke engines consists of 25 or 30 parts, most of which require precision machining and heat treating. These parts add as much as $50 to the retail price, compromise reliability, and complicate repairs. As Henry Ford used to say, "Parts you don't have, can't break."

Two-cycle lubrication is automatic, once the fuel is mixed. But four-cycle engines need frequent oil changes and level checks, chores that contractors cannot depend upon their field personnel to perform. In addition, there are reports of problems with these new engines when operated off the horizontal.

But the advent of hybrids, which combine some of the best features of two-strokes with superior fuel economy and less pollution, shifts the balance.

Hybrids

The STIHL 4-Mix®used to power several of the company's garden tools, combines crankcase induction with four-stroke operation. The dry crankcase permits the engine to run at any angle and eliminates worries about oil levels and change intervals. And the four-stroke cycle, which allocates a full piston stroke to scavenging, significantly reduces HC emissions without the need for a catalytic converter. As a bonus, 4-Mix engines deliver 30% better fuel economy and 14% more torque than equivalent two-strokes.

Figure 1-12 shows the general layout. The upper crankcase, cylinder barrel, and head are integrated into a single aluminum casting. Overhead valves operate from a single cam lobe, a feature that reduces engine size and weight (Fig. 1-13). The cam also incorporates an automatic compression release.

Oiling is accomplished by patented, two-stage process. As the piston rises toward TDC, it creates a partial vacuum in the crankcase (Fig. 1-14). The fresh charge, consisting of a 50:1 mix of fuel and lube oil, enters through the inlet pipe (1) and bypass port (2) and passes over the cylinder head on its way to the crankcase. En route, the fuel charge wets all moving parts, including the critical camshaft lever.

Cleaner two-cycle exhaust

Survival of the two-stroke engine depends upon finding ways to limit HC and CO exhaust emissions. Currently two approaches are used: stratified charging and exhaust after treatment.

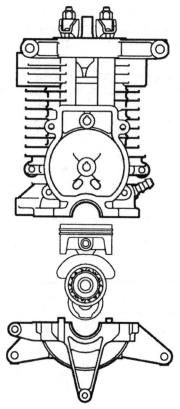

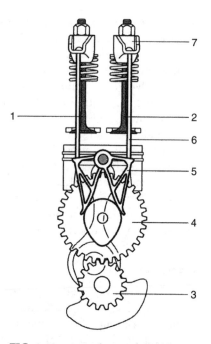

FIG. 1-12. *The horizontally split crankcase is a welcome feature that provides easy access to rotating components.*

FIG. 1-13. *A single cam lobe (4), driven at half speed through the crankshaft gear (3) and through a pivoted lever (5), pushrods (6), and rocker arms (7) actuates the exhaust (1) and inlet valves (2). Generously sized drive gears, coupled with anti-friction bearings, are designed for durability.*

Stratified scavenging

One way to sanitize two-strokes is to purge the cylinder with air before introducing fuel. As developed by RedMax-Komatsu Zenoah, stratified scavenging employs a Walbro WYA double-barreled carburetor, with one bore for air delivery and the other for the air/fuel mixture (Fig. 1-16).

The RedMax Stage 1 incorporates two transfer ports, each of which is fitted with a reed valve that opens to the air-only carburetor bore. An additional reed valve connects the second, or fueling, carburetor bore with the crankcase.

As the piston uncovers the exhaust port, residual gases blow down and leave a vacuum behind in the cylinder. Transfer-port reed valves open to admit purge air from the carburetor. Meanwhile the falling piston builds pressure in the crankcase and in the transfer ports. Within 10° or so of crankshaft rotation after blowdown, sufficient pressure is developed to close the

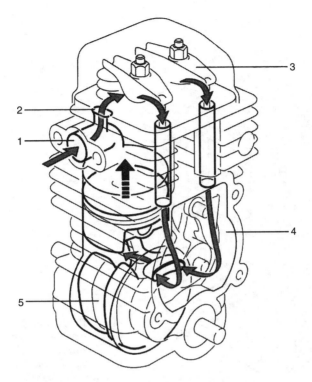

FIG. 1-14. *The 4-Mix® draws fuel into the crankcase through the bypass port as the piston rises. The downward movement of the piston pressurizes the crankcase. Oil-impregnated fuel then leaves the crankcase, lubricating the critical valve-gear parts for a second time (Fig. 1-15).*

transfer-port reed valves. Air no longer enters the cylinder, which then fills with the fresh mixture from the crankcase.

According to RedMax, stratified scavenging reduces HC emissions by as much as 80% and, under optimum conditions, boosts thermal efficiency by 51%. On the other hand, cylinder filling suffers because some scavenge air remains in the cylinder after the exhaust port closes. The prototype version developed slightly less power than a reference engine with conventional porting. NOx emissions, fueled by the additional oxygen, showed a small increase. In 1998 the RedMax became the first production two-stroke to receive CARB Tier-2 certification. That year also saw the certification of a proof-of-concept engine, the result of a $10-million collaboration by Tanaka and CARB. Unlike the EPA, California regulators take an active part in the development of technology.

Surprisingly enough, the Strato-Charged RedMax encountered buyer resistance. Potential buyers interpreted the unobtrusive exhaust note as evidence of lack of power. But reduced fuel consumption, on the order of 34%,

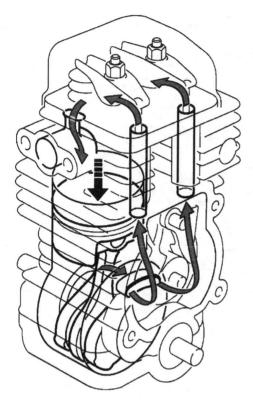

FIG. 1-15. *As the piston falls, the pressurized fuel mixture flows out of the crankcase and back to the intake pipe for induction into the cylinder. In the process, moving parts are again wetted.*

FIG. 1-16. *The Walbro WYA carburetor features a 12-mm bore, an adjustable idle jet, and can be fitted with an accelerator pump. While designed for stratified-charge two-stroke engines, this carburetor can also be adapted to four-strokes.*

made converts. The big breakthrough came in 1999, when STIHL purchased 60,000 of the engines for the California market.

Catalytic converters

The CARB Tier 2 Tanaka Pure-Fire 260PF trimmer/brush cutter employs a catalytic muffler and a unique induction system. The engine has a 33-mm bore and a short 28-mm stroke for a displacement of 24 cc. With a dry weight of only 12.4 lb and a power output of 1.3 hp at 11,000 rpm, the 260PF has one of the best power-to-weight ratios in the industry.

At first glance, the PureFire could pass for any other piston-ported engine. Air and fuel are drawn into the crankcase during the upstroke and forced out through transfer ports on the downstroke. Fuel and air move out of the crankcase and into a channel that leads to the transfer ports. A small-diameter orifice connects the crankcase cavity with the transfer-port feed. This orifice gives velocity to the charge, which absorbs heat in its journey to the transfer ports. Velocity and heat assist in fuel atomization, a precondition for efficient combustion. The photo on the cover of this book illustrates Purefire gas flow.

HC and CO emissions that survive are oxidized in the catalytic muffler. The catalyst consists of a thin veneer of platinum, palladium, or rhodium applied over a ceramic or metal matrix. A two-way converter, that is., one that oxidizes HC and CO, is all that's needed. NOx emissions do not present much of a problem for carbureted two-strokes.

In the presence of heat, the converter oxidizes hydrocarbons into carbon dioxide and water. The basic reaction is

$$HC + O_2 \rightarrow CO_2 + H_2O.$$

This exothermic reaction boosts exhaust temperatures to 1000°C and higher. Carbon monoxide combines with oxygen to form carbon dioxide:

$$CO + 1/2O_2 \rightarrow CO_2.$$

Other reactions involving CO release water and hydrogen.

The first version of PureFire technology—the 250PF—relied upon the converter to scrub 40% of exhaust emissions. The recently introduced 260PF cuts the percentage to 30% through a redesign of the combustion chamber. Consequently, the converter runs cooler and should last longer.

A catalytic converter intensifies maintenance requirements. Rich mixtures resulting from improper carburetor adjustments or dirty air filters must be avoided. Converter burnout occurs at 1400°C or about 300°C more than normal for high-output engines. Excessive oil or the wrong type of oil fouls the catalyst. And finally, silicone adhesives or carburetor cleaners containing lead, phosphorous, or silicone must not be used. But, even with these precautions, converters are sacrificial items that need periodic replacement.

While I do not have figures for handheld equipment, the Indian government estimates that catalytic converters for light motorcycles have a useful life of only 15,000 km.

Marks of quality

Designers of small motorcycle engines are a conservative bunch and rarely deviate from a pattern that was established during the 1930s. Almost without exception, these engines have vertically split crankcases, ball main bearings, one-piece connecting rods, and loop-scavenged cylinders. Since design is frozen, variations in quality come down to the choice of materials and the precision of manufacture. Japanese and European makes are the obvious leaders, with Chinese and East Indian bikes trailing far behind. While many readers might protest, the EPA calculates the useful life of small motorcycles (those displacing between 150 and 169 cc) at 12,000 km (7456 mi).

Designers of handheld equipment are much more flexible, in part, because the machines are tailored to very different markets. It's possible to buy a chainsaw for $150 or weed trimmer for half that. But the money saved will soon be spent on repairs. Midrange handhelds cost, on the average, around $300 per unit. Expect to pay $600 or more for professional quality two-stroke equipment. In no particular order, the top brands include names such as STIHL, RedMax, top-line Echos, Shindaiwa, Tanaka, Husqvarna, Dolmar, and Jonsered. Find out what local landscape contractors use, since quality involves more than the brand name. You want a machine from a nearby dealer, who has a good parts inventory and who honors warranty claims.

What is the useful life of handheld equipment? Manufacturers are coy about this question, but we do know that the design life—the life engineers try for—is 1000 hours for professional quality tools. Echo insists that their engines, which are not dramatically different than other upper-echelon products, have a design life of 1200 to 1500 hours or more. In the writer's experience, low-end, mass-marketed tools can self-destruct in as little as 50 hours.

The factory warranty period reflects how well the design life goals are achieved. Good-quality garden and construction tools carry a warranty of 2 years in private use, 1 year commercial, and 90 days rental. Some offer longer periods and certain components such as ignition modules and drive-line parts may be covered for the life of the equipment. Chainsaw warranties typically extend for a year of private use.

Another indication of the way manufacturers judge the longevity of their products has come about because the EPA insists that exhaust emissions be tested twice, once when the engine is new and a second time at the end of its useful life. According to the rule, published in 40 CFR, Chapter 1, 90.105, manufacturers must certify the time of the second test. The choice is between 125, 250, and 500 operating hours. Useful life is calculated on the basis of engineering evaluations of wear, surveys of engines in the field, customer complaints and other hard data. No manufacturer has opted for the 500-hour test.

Design features to look for are

- Two main bearings, one on each end of the crankshaft. Cheaper and almost certainly less durable engines use crankshafts cantilevered off of a single bearing (Fig. 1-17).
- High-quality engines, such as the STIHL, RedMax, and Tanaka employ replaceable crankshaft seals that are more reliable and easier to replace than bearings with integrated seals (Fig. 1-18).
- Chrome-plated or Nikasil-coated cylinder bore. Some manufacturers merely chrome the piston.
- Stress-relieved cylinders. Tanaka and a few other manufacturers allow cylinders to age before final machining.
- Thin, closely packed cylinder fins for efficient cooling.
- Four cylinder-to-crankcase bolts, rather than two (Fig. 1-18). Two hold-down bolts are a proscription for air leaks, although it must be said that the RedMax two-bolt arrangement with locking tabs gives no trouble.
- Two compression rings. Low-end products get by with one.

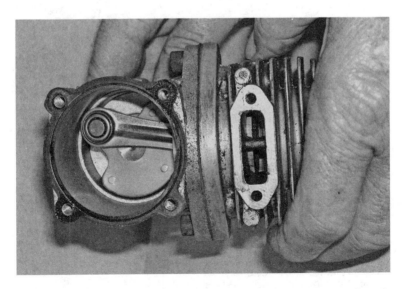

FIG. 1-17. *Hanging the crankshaft off a single main bearing dramatically reduces costs. The vestigial crankshaft requires little machining and the crankcase cover can be made of plastic.* Robert Shelby

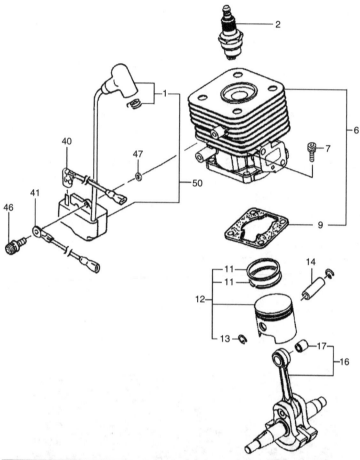

Item	Part number	Description	Qty	Comments
1-1	1570168090	Cap, spark plug Ass'y	1	
1-2	01802233200	Plug, spark	1	
1-6	0010760090	Cylinder set	1	Includes gasket
1-7	99461050184	Bolt, hex, 5 X 18, S	4	
1-9	0170735020	Gasket, cylinder	1	
1-11	04101601200	Ring, piston	2	
1-12	0300750090	Piston set	1	Includes rings and circlips
1-13	03901620200	Circlip, piston pin	2	
1-14	0370704520	Pin, piston	1	
1-16	0460735082	Crankshaft	1	
1-17	99962081225	Bearing, needle, F-810	1	
1-40	1780168080	Cord, stop	1	
1-41	2740168080	Cord, earth	1	
1-46	99464040184	Bolt, hext 4X18WS	2	
1-47	1820186A201	Washer, special	2	
1-50	1672101290	Coil, ingition	1	

FIG. 1-18. *Tanaka 260 PF is an example of design quality.* Hitachi Koki USA

- Wrist pin with a caged roller bearing and thrust washers. Few, if any, low-priced engines incorporate thrust washers. One manufacturer goes so far as to eliminate the circlips that prevent the wrist pin from walking into the cylinder bore.
- One-piece connecting rod as opposed to a sheet-steel stamping.
- A flywheel massive enough to smooth compression resistance during cranking.
- A digital ignition module that eliminates kickbacks during starting.

There is also the question of how power is transmitted through 90°, which is a requirement for most handheld tools. The easy way is to make the transition is mount the cutter head on a curved shaft. A flexible cable, a kind of jumbo speedometer cable, transmits power. The cable whips and quickly destroys its bushings. Quality machines employ a solid steel drive shaft, about 7 mm in diameter, and bevel gears at the cutter head. The ends of the shaft should have proper splines (D-shaped or squared ends need not apply) and ride on ball bearings. Oil-lite bushings are a distant second choice and nylon bushings are hardly worth mentioning.

But be aware: Some midmarket trimmers have the prerequisite gear box, but drive through a flexible cable.

The learning curve

Things learned from working on automobiles and four-stroke lawnmowers do not always apply to two-strokes. This is why small-engine shops usually have a resident two-stroke guru, who can be called on when simple fixes fail.

The basic problem is getting rid of the heat generated by engines that fire every revolution. Catalytic converters add to the heat load as does the scale effect. Because of reduced surface area, the smaller the engine, the more heat retained.

Combine heat with the high oil consumption of two-cycle engines and we have a recipe for carbon production. After long use, carbon deposits in the exhaust port and muffler throttle the engine. Carbon also collects in the ring grooves, where it robs the rings of their elasticity. If rings cannot flex, they cannot seal. Large fragments of carbon, embryonic diamonds, can break off and score the bore and piston. Most scoring occurs in areas adjacent to the exhaust port, which runs hot and sees little by way of lubrication. As a point of interest, the bridge—the knife-like rib running vertically between the exhaust ports—is the hottest part of the engine. But without a bridge, the rings would expand to fill the port and snag.

Heat also distorts plastic shrouding, causing air leaks and converting fuel lines into something resembling uncooked spaghetti. Compromised fuel lines can leak fuel or air.

Another factor to be aware of is the sensitivity of the cylinder/ring seal. Wear and scuffing that go unnoticed on larger engines take on major importance when cylinder diameters are reduced to 30 or 35 mm. The ring seal also pressurizes the crankcase, a function that becomes critical during starting, when rpm is low. Consequently a small two-stroke needs a minimum of 90 psi cranking compression. A four-stroke can be persuaded to run with 60 psi. Crankcase compression also depends upon the integrity of the crankshaft seals and, when fitted, the reed valve.

And finally, there is the matter of carburetors. In order to operate at large angles off the horizontal, two-stroke engines have diaphragm carburetors with integral fuel pumps. These carburetors are a world apart from the float-type units found on most four-cycle engines.

To summarize, the differences between two-stroke gurus and ordinary mechanics are that gurus appreciate the importance of cylinder and crankcase sealing, check that rings are free to flex, look on fuel lines with suspicion, and have experience with diaphragm carburetors.

Long-term storage

At least half of the problems associated with two-cycle engines come about because of improper storage. Before putting an engine up for any extended period,

- Remove the shrouding and clean the cylinder fins with a brush and solvent.
- Remove debris from the cutter head, blade, and drive sprocket.
- Inspect and, if necessary, clean the flame arrestor screen that will either be inside the muffler or clipped to it.
- Drain the fuel tank and run the engine *at idle* until it stops. Do not open the throttle, since running dry of fuel also robs the engine of lubrication.
- Remove the spark plug and pour a teaspoon or so of oil into the cylinder. Slowly turn the engine over several revolutions to distribute the oil. Replace the spark plug and reconnect the ignition lead.
- Safely discard fuel that will not be used during the next 3 to 4 weeks. Inspect the fuel container for rust or varnish build up.

Safety

By their nature, gasoline engines pose risks. The fuel is highly volatile, mufflers and catalytic converters run hot, exposed parts revolve at high speeds, and, when confined in a closed space, the exhaust can be lethal. These dangers are exacerbated when the engine is handheld or strapped to one's back.

Gasoline

A U.S. gallon of gasoline contains 125,000 Btu, which represents more heat energy than an equivalent weight of TNT. At temperatures above 45°F (7.2°C), open or leaking gasoline containers give off vapor. Should it find an ignition source, the vapor trail acts as a fuse to convey the flame back to the container.

Store fuel in an approved safety container—not a plastic jug. The container should be tightly sealed and filled to no more 90% of capacity to allow for heat expansion. Gasoline and other flammables should be stored in a locker, remote from the shop and house.

Allow at least 5 minutes for the engine to cool before refueling or opening a fuel line. If working outdoors is impractical, make certain the shop is well ventilated and that no ignition sources are present. Obviously one would not weld, grind, or smoke in the presence of fuel vapor, but one should also be aware that any spark, as from switching on a light, can become an ignition source. Before starting an engine after adding fuel, wipe up any spills, allow ample time for evaporation, and position the fuel container at least 10 ft (3 m) from the machine.

Although many mechanics prime engines by injecting gasoline into the carburetor or cylinder, many of the same mechanics are admitted to hospital emergency rooms for burn treatments. When an engine must be primed to verify fuel delivery, use aerosol carburetor cleaner, never gasoline. Work outside, spray a tiny amount of cleaner into the air intake, and replace the air filter element before starting. The air filter acts as a spark arrestor.

Do not use gasoline as a cleaning fluid. A heavier hydrocarbon, such as kerosene or "safety solvent" works about as well and eliminates much of the fire and exposure hazard. Gunk, available in aerosol cans from auto parts houses, does an excellent job of removing grease. But no solvent is entirely free of hazard; California health authorities classify Gunk as a carcinogen.

And finally, equip the shop with at least one Class B fire extinguisher. Water merely spreads gasoline and oil fires.

Carbon monoxide

Never operate an engine in an enclosed space. When inhaled, carbon monoxide (CO) displaces oxygen in the bloodstream. Low levels of exposure result in nausea and headaches; higher levels are lethal.

Rotating parts

The business end of cutting and drilling tools present obvious hazards, and string trimmers, some of which turn 11,000 rpm, are by no means innocuous. Complement the safety instructions in the owner's manual—which can never be complete—with common sense.

Be aware of the possibility of accidental starting when carrying out service work. An engine secured in a vise by its crankshaft can, and sometimes does, start when rotated. If the shrouding is in place, the spark plug will not be accessible to a hammer; about all one can do is to clear the area and wait for the tank to run dry.

Open sparks

While we routinely test for ignition voltage by arcing the high-tension lead to ground, this procedure entails risk. The safer approach is to use a shielded ignition tester, as described in the "Troubleshooting" and "Ignition systems" chapters. These testers confine the spark behind a transparent window.

Recalls

The U.S. Consumer Products Safety Commission website at www.cpsc.gov performs a valuable service by alerting mechanics and owners to safety hazards that, because they are statistical in nature, might otherwise go unnoticed. Chainsaws have been recalled because of fuel leaks and flywheel disintegration. The muffler support strap on RedMax EB 6200, EB 7000, and EB 7001 backpack blowers has, on several occasions, torn away from the muffler, leaving a hole that diverts hot exhaust gases toward the plastic fuel tank.

Conclusion

Now that we understand how these engines work and something about the forces that have shaped them, it's time break out the tools and get to work. The next chapter deals with troubleshooting, the critical skill that separates real mechanics from part changers.

2

Troubleshooting

Few experiences are as frustrating as an engine that refuses to start or that runs a few seconds and dies. One continues to crank, hoping that somehow things will fix themselves. We all do this, even those of us who should know better. If it's any consolation, porpoises exhibit the same sort of magical thinking. At a naval research facility in Maryland the animals were routinely given slight electrical shocks, which they found pleasurable. One day the generator failed and the porpoises gathered round, whistling. When the recording of their sounds was played at slow speeds, the porpoises were saying, "Testing, one, two, three." They were cranking, trying to get the generator started.

A two-cycle engine consists of four systems:

- Mechanical—piston, crankshaft, and connecting rod
- Ignition—spark generator and spark plug
- Fuel—carburetor, filter, and associated plumbing
- Starting—rewind starter

When something goes wrong, we must first decide upon the system involved. How to relate malfunctions to engine systems is the subject of this chapter. Once categorized, repairs are simple.

But understand that engines, even little two-strokes, are complicated enough to elude formula. Troubleshooting charts and logic trees oversimplify because they frame diagnostic questions in such a way that the answers can only be "Yes" or "No." In the real world answers are rarely so pat. We cannot see air leaking past a gasket or the failure of the spark plug to fire when screwed into the cylinder. But we can assume that a torn gasket leaks and that a dirty spark plug misfires under compression.

In other words, troubleshooting is a gambling game—we are guided by probabilities and wager time, parts and ego on the outcome. Odds improve with close observation, conditioned by an understanding of the importance of spark, fuel delivery, and compression.

Things to keep in mind

An engine must run if it has

- Sufficient spark to fire the mixture at cranking speed. We can determine the presence of spark with an ignition tester but, short of laboratory instruments, we cannot measure spark voltage or energy.
- Fuel in the right amounts to support combustion. Too much fuel is as bad as too little. The condition of the spark-plug tip is our gauge of fuel delivery.
- Adequate compression. Small two-stroke engines need a minimum of 90 psi to start.
- Free-flowing exhaust. Carbon accumulations in the exhaust ports and muffler rarely prevent an engine from starting, but drastically reduce power outputs.

Tools and supplies

Two-strokes are held together with metric fasteners, star-pattern Torx screws, and an occasional Phillips or Allen screw. Quarter-inch drive sockets are adequate for everything except the flywheel nut. U.S. standard 1/2-in. wrenches fit (somewhat loosely) 13-mm bolts and 9/16-in. wrenches work for 14 mm. You will also need a spark-plug socket wrench, an ignition tester, and a compression gauge.

Because they are torque-limiting, that is, designed to slip, Phillips screws present a problem. Striking the screwdriver with a hammer sometimes helps, but the ultimate persuader is a hammer-impact wrench. Once you get the screws out, replace them with Allen or Torx fasteners.

Supplies include fresh fuel, mixed with the correct amount of two-stroke oil, solvent, aerosol carburetor cleaner, and lint-free rags. Automotive brake cleaner is a potent, but toxic, degreaser that attacks paint and some plastics.

And a word of caution: no amount of care can eliminate gasoline spills associated with draining fuel tanks and disconnecting fuel lines. Ideally, the work should be done outside. If that is impractical, the work area should be well-ventilated and remote from pilot lights and other ignition sources. Wipe up spills promptly and allow plenty of time for traces of gasoline to evaporate before cranking the engine. Keep a fire extinguisher handy.

Another tool, often overlooked, is a notebook small enough to carry in a top pocket. Much frustration will be avoided by recording where fasteners go, which ones were treated with sealant, how fuel-system plumbing is routed, and other details that seem so obvious upon disassembly and so mysterious when it's time to put things back together.

Preliminaries

When an engine breaks down in the field, we try to zero in on the problem, fix it, and go on about our business. In a repair shop, the urgency is usually less, and we can take the time to round up the usual aspects by replacing parts most likely to have failed.

- *Inspection* Check for obvious faults. These include fuel leaks, weak compression, loose carburetor, crankshaft binds, and oil stains adjacent to gasket surfaces. If working on someone else's engine, bolts that show fresh wrench marks mean repairs were attempted, probably since the engine last ran. Such abortive attempts at repair complicate matters since you must now identify the original problem and whatever the hapless mechanic might have done wrong.
- *Spark plug* Begin by replacing the spark plug with a new, or known-good, plug of the type recommended in the operator's manual. This action will correct 90% of engine malfunctions. Gap to specification, which for most handheld engines is 0.016 in. and for larger units 0.025 in. (Fig. 2-1). A cam-type gauge gives less ambiguous readings than a conventional, flat-blade feeler gauge (Fig. 2-2).
- *Fuel* If there is any question about fuel quality, take the machine outdoors and drain the contents of the tank into a glass container. An acrid smell, a brownish taint, or globules of water in bottom of the container mean that the fuel system has been contaminated. Flush the tank with hot water and detergent, replace the fuel filter, and overhaul the carburetor with new diaphragms, inlet needle, and gaskets as described in Chap. 4.
- *Fuel filter* Replace the filter, using the correct factory part. "Universal" automotive filters clog quickly when used on gravity-fed small engines. Use a hooked wire to retrieve in-tank filters of the kind found on handheld equipment (Fig. 2-3).
- *Air filter* Most two-cycle engines use a polyurethane foam filter that should be cleaned with soapy water and lightly oiled with a proprietary lubricant. The foam filter may be backed up with a pleated paper filter, which is a sacrificial element requiring routine replacement. You may also encounter a third filter downstream of the paper element. This filter is a backstop to prevent dust entry when servicing primary filters.

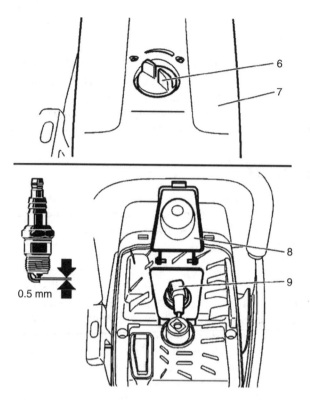

FIG. 2-1. *Narrow spark-plug gaps are the norm for small engines. In the case of this Wacker Neuson chainsaw, the 0.5-mm gap translates as 0.013 in.*

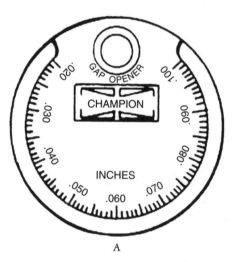

A

FIG. 2-2 *A Champion taper gap gauge, available in auto parts stores or from Tecumseh as PN 670256, can be used to widen the gap (A). To reduce the clearance, tap the side electrode against a hard surface (B).* Bosch

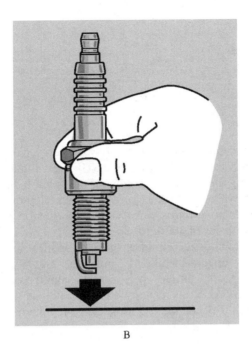

B

FIG. 2-2 *(Continued)*

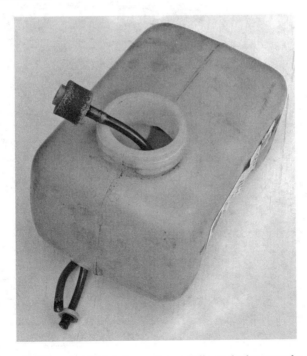

FIG. 2-3 *In-tank filters should be changed annually and whenever the equipment comes into the shop.* Robert Shelby

Tests

Diagnosis begins with a series of tests and observations.

Spark output

Anyone who contemplates doing much work on small engines should invest in a Briggs & Stratton PN19053 or a STIHL Zat 4 ignition tester. These tools eliminate any fire hazard by confining the spark behind a transparent window. People with pacemakers or other medical problems should not expose themselves to electroshock. Solid-state ignition systems produce something on the order of 30 kV, open-circuit.

That said, the traditional method of verifying spark delivery is to hold the spark plug against engine metal and crank (Fig. 2-4). So long as the plug makes firm contact, voltage goes to ground through the engine and not the mechanic. A variation on this technique is illustrated in Fig. 2-5.

The spark should occur once during every revolution, as regular as a heart beat. No spark means a

FIG. 2-4. *This method of checking spark leaves much to be desired. Plastic engine covers must be removed to access ground and unshielded spark presents a fire hazard.* Robert Shelby

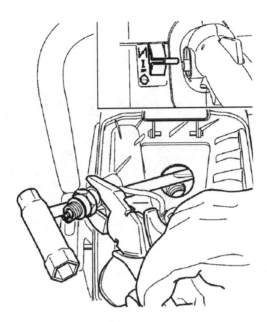

FIG. 2-5. *With a spark-plug wrench inserted into the spark port, ground the spark plug against the wrench. Note that the wrench handle must not extend into the combustion chamber, if engine damage is to be avoided. The mechanic has also taken the precaution of using insulated pliers.* Wacker Neuson

- Sheared or deformed flywheel key.
- A short in the wiring going to the kill switch or another problem with the primary circuit that sometimes includes safety interlocks, which are notoriously unreliable. Test the integrity of this circuit by cranking with the kill-switch wiring disconnected from the ignition module (Fig. 2-6). Do the same for lights and other electrical components that may be present.
- *Ignition module failure* Solid-state modules, which have been standard on small engines for almost 30 years, are not repairable or, for that matter, testable in any meaningful sense. Nor are electrical parts returnable after purchase. If you suspect that the module had failed, the only option is to test by substitution of a new, or known-good, unit.
- *"Whiskers"* While this phenomenon is rare, it's worth mentioning. Two-cycle spark plugs, even brand new spark plugs, can spontaneously develop "whiskers"—fine, hairlike growths between the side and center electrodes that short voltage to ground. In the wake of a whisker-induced ultra-light aircraft accident in England, Rotax replaced the 532 engine, which used a single spark plug per cylinder, the dual-plug 582.

An ignition tester is also useful when diagnosing engines that run for a few minutes and then shut themselves off. Connect the tester to the high-tension

FIG. 2-6. *The screwdriver points to connection on the ignition coil for the kill switch.*
Robert Shelby

lead, in series with the spark plug. Start the engine and watch the spark as the engine dies and coasts to a stop. Ignition should continue until the flywheel almost stops revolving. If ignition ceases before shutdown, the ignition module has failed, probably because of heat. If ignition remains constant, then the source of the problem lies elsewhere, probably in the fuel system.

Fuel system

Fuel systems go wrong by delivering too little or too much fuel.

No fuel   Crank a cold engine five or six times and remove the spark plug. If fuel is entering the cylinder, the tip of the spark plug will be wetted and smell of gasoline. If the plug is dry, momentarily disconnect the fuel line to the carburetor. Unless there is an obstruction upstream—tank filter, screen, or clogged line—fuel should dribble out of the open line. Repair as necessary.

When fuel is present at the carburetor inlet, the logical response is to bypass the carburetor. Remove the spark plug and spray a small amount of carburetor cleaner into the cylinder. Reinstall the spark plug and crank. If the engine does not start, the problem is almost certainly the spark plug or ignition module. If it starts and runs a few seconds, we can assume that fuel starvation is the problem. Some malfunction prevents fuel from passing through the carburetor and into the combustion chamber.

To further isolate the problem, remove the air-filter element and spray carburetor cleaner into the air horn. Replace the filter.

Warning: Do not run an engine without the air-filter in place.

If the engine runs until the aerosol is consumed, we can be confident that things downstream of the carburetor are normal. That is, the reed valve functions and the piston seals well enough to draw fuel into the crankcase. If the crankcase seals leak, the leak does not inhibit starting. Since we have already verified that fuel is available to the carburetor, the problem must be in the carburetor itself.

Chapter 4 describes how to make the necessary repairs.

Too much fuel   Flooding a cold engine soaks the spark plug with raw fuel and, in more extreme cases, sends fuel dribbling out the carburetor air horn. Flooding can be induced by heroic cranking, but more often is the fault of the carburetor. Float-type carburetors flood because of dirt between the inlet needle and seat or because of a hung float. Diaphragm carburetors are less prone to flooding, but will do so if the inlet needle sticks in the open position.

Flooding in a hot engine is more difficult to detect, since the fuel vaporizes. Sometimes you can see a curl of vapor leave the spark-plug port as the plug is removed. In many cases, hot flooding comes about because of heroic cranking or because the choke was left engaged. If the problem persists when the engine has cooled down, assume that the carburetor is at fault.

Lean mixtures   Lean mixtures are a cause for alarm. The operator might notice flat spots during acceeration or that the engine runs better with a partially closed choke. A mechanic will see the bleached, bone-white spark plug insulator (Fig. 2-7) and immediately suspect piston damage of the sort pictured in Fig. 2-8. In extreme cases, high combustion temperatures melt the piston crown.

FIG. 2-7. *A white insulator tip is the first symptom of overheating, almost always caused by lean mixtures. Too much ignition advance on engines with adjustable timing has the same effect. The damage shown in the photo comes about when overheating results in detonation or preignition.* U.S. Bosch

FIG. 2-8. *High combustion temperatures are death on pistons. This example exhibits pock marks left by metal fragments, deep scores, and, near the crown on the right of the photo, weld splotches.* Robert Shelby

Lean mixtures are attributable to

- Carburetor malfunctions (Chap. 4)
- Leaking crankshaft seals (Chap. 6)
- Leaks at the carburetor flange, (Fig. 2-9)

Rich mixtures   Black, furry deposits that you can wipe off with your fingers are evidence of rich combustion (Fig. 2-10). The engine is drowning in gasoline. Suspect a maladjusted carburetor and/or a dirty air filter. Unlike lean mixtures, gasoline-rich mixtures do no permanent damage, at least on engines without catalytic converters.

It is also true that chronic misfiring darkens the spark plug. A replacement plug usually solves the problem.

Compression   Two-cycle engines develop two kinds of compression: crankcase and cylinder. Weak crankcase compression reveals itself as a refusal to ingest carburetor cleaner sprayed into the air intake. Some mechanics claim to be able to sense the crankcase pressure, which never amounts to more

FIG. 2-9. *Carburetors often vibrate loose from their mounts and leak air.* Robert Shelby

FIG. 2-10. *The darker the color of the insulator, the richer the mixture.* U.S. Bosch

than 5 or 6 psi, as resistance on starter cord. The rest of us test crankcase integrity as described in Chap. 6.

Cylinder compression is another matter and should be checked whenever an engine enters the shop. Vintage engines—West Bends, Clintons, Power Products, Villiers—could sometimes be persuaded to start with as little as 60 psi of cylinder pressure. Modern engines require 90 psi or more.

FIG. 2-11. *Cylinder compression is a basic indication of engine health.* Robert Shelby

Thread a compression gauge in the spark-plug port, open the throttle and choke, and crank three times (Fig. 2-11).

If compression is a problem, remove the muffler and, using a soft probe, verify if the rings move in and out of their grooves. Two-cycle rings are pegged to prevent rotation, but should move inward a few thousandths of an inch and spring back when pressure on them is relieved. Stuck or broken rings cannot seal.

Now rotate the flywheel to bring the piston below the exhaust port. Using a penlight, examine the far cylinder wall for scores. Any scoring costs compression. Raise the piston above the port. Scores, weld splotches, or extensive discoloration of the piston skirt are fatal (Fig. 2-12).

Piston damage, which nearly always involves collateral damage to the cylinder, may signal that it's time to fold. Factory short blocks, that is, new rotating parts installed in new cylinder and crankcase castings, usually cost more than the engine is worth.

When the engine spins freely without any resistance on the starter cord, we have zero compression. This condition can be caused by a holed piston or, as more often is the case, by a thrown rod (Fig. 2-13). Repair is rarely worthwhile.

Backpressure   Although "backpressure" is a misnomer, the term describes what happens when the plumbing clogs with carbon. The engine chokes on its own exhaust.

FIG. 2-12. *What one hopes not to see when peering through the exhaust port.* Robert Shelby

FIG. 2-13. *A parted rod, resulting from big-end bearing failure.* Robert Shelby

Lower the piston and, using a soft brass or copper tool, scrape the exhaust ports clean (Fig. 2-14). Clear the cylinder of carbon fragments by spinning the flywheel rapidly with the spark plug removed.

Chain saws and many garden tools incorporate a spark arrestor in the form of a screen over the exhaust ports or inside the muffler (Fig. 2-15). Gently clean

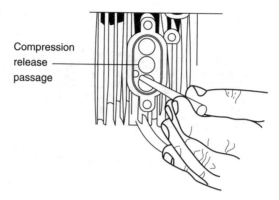

Compression
release
passage

FIG. 2-14. *Some Tecumseh engines have a compression-release port, which also requires cleaning.*

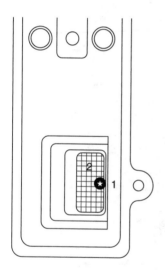

FIG. 2-15. *Most handheld and agricultural engines include a spark arrestor similar to the STIHL unit shown. Clean the screen in solvent, being careful not to damage the fragile mesh. Torn screens should be replaced.*

the screen with a brass wire brush. Noncatalytic mufflers can be cleaned by immersion for an hour or so in hot water and caustic soda. Clogged catalytic converters must be replaced.

Complaints

Now that we have an appreciation of what engines require, complaints about engine behavior are not difficult to resolve.

No start

Failure to start with a new spark plug, fuel filter, and a quarter tank of fresh premix can have several causes, most of them obvious. We can pretty well forget about carburetor adjustments (unless someone has jiggered with the carburetor since the engine ran last), carbonized exhaust ports, and other subtleties.

Check ignition output as described under "Spark Output" earlier. No spark usually means a failed flywheel key, although the wire going to the kill switch should be checked for frayed insulation or burnt spots.

Crank the engine to determine if fuel gets into the cylinder. To verify, spray aerosol carburetor cleaner into the cylinder and carburetor intake, as described under "No fuel" earlier. If the engine runs on aerosol, but not on gasoline, we can be confident that the problem is in the fuel system. Likely culprits are a dirty carburetor, hardened or leaking diaphragms, or a fuel-pump malfunction. These and other fuel system malfunctions are detailed in Chap. 4.

And finally make a compression test, looking to find that magic 90-psi number.

Reluctance to start

That the engine starts at all means that compression and the ignition module are probably okay. But check the compression and, if you have not already done so, replace the spark plug with a known good one. The most likely culprit is insufficient fuel, especially if the engine is of recent vintage. Clean and use whatever carburetor adjustments are available to richen the mixture. In some cases, the only recourse is to fit an earlier model, fully adjustable carburetor.

Runs a few minutes and quits

Intermittent malfunctions are, almost without exception, confined to the ignition and fuel systems. Test ignition output with the engine running is described under "Spark Output." If spark persists during coastdown, the problem lies in the fuel system. Suspect a clogged fuel-tank vent (Fig. 2-16) or a sticking carburetor inlet needle.

Lack of power

Our perceptions of power are subjective. When the complaint is real, the first priority is to distinguish between the engine and what it drives. The fault may be a dull saw chain, an oil-soaked centrifugal clutch, dragging bearings, or an overtightened belt.

Air vent hole   Air vent valve

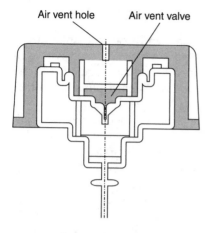

FIG. 2-16. *Fuel-tank vents for handheld engines can be fairly complicated, as shown on this Tanaka unit. Some mechanics ignore the very real fire hazard and check vent operation by running the engine with the cap loose.*

If power loss is the fault of the engine, the condition usually develops over time, due to the slow degradation of all systems. About all one can do is to rebuild the carburetor, install new rings, and verify that the exhaust system flows freely. A slightly richer than normal mixture may help.

Refusal to idle or to run at high speed

These carburetor-related problems are discussed in Chap. 4.

Flywheel binds

Disconnect the engine from its driven element and retest. If the flywheel continues to bind, the problem is serious. See Chap. 6 for repair instructions.

Rewind starter fails to engage

See Chap. 5.

3

Ignition systems

Any small-engine ignition, even the most primitive Wico magneto, can generate the 12 kV needed to fire a clean spark plug. But none of these systems deliver the voltage needed to fire dirty plugs or ignite less-than-optimum air-fuel mixtures. This is why small engines are so susceptible to flooding and why spark plugs need such frequent replacement.

Automotive ignitions put out about 40 kV and the spark plugs are gapped accordingly, with 0.050 in. as the norm. No small-engine spark plugs are gapped more than 0.025 in. and many run as narrow as 0.008 in.

More voltage would make starting easier and, in combination with a wider spark gap, would smooth combustion. In-cylinder mixtures are not homogeneous; in some areas the mixture is too lean to ignite or, when ignited, burns slowly. The pressure rise in the cylinder becomes erratic, as if ignition timing were wandering. A wide spark gap betters the odds of encountering a rich, combustion-prone mixture.

That said, we pretty much have to live with OEM ignition systems. About all that one can do is to keep several spark plugs on hand and accept that small engines are prone to flooding. Once this happens, stop cranking, let the machine sit for half an hour, and it usually will start.

It is possible to update transistorized and point-and-condenser magnetos with capacitive discharge (CD) modules, but the exercise would be prohibitive for anyone who does not have a stock of used parts. It is also possible to install a modern automotive ignition system. Brian Miller (http://gardentrac-torpullingtips.com/a1elect.htm) can supply the parts and the expertise. The conversion requires a 12V source of voltage, which can be a problem. MSD also makes a battery-powered capacitive discharge ignition (CDI) for small engines, but for reasons that are not clear, limits applications to four-strokes.

Diagnosis

Solid-state ignition modules behave like light bulbs. That is, they usually work fine until complete failure. Old-time magnetos often failed by degrees as contact points oxidized and became resistive.

As pointed out in the previous chapter, open sparks and a gasoline engine are not a good combination. Some gasoline vapor will be present, especially if the test is done with the spark plug removed from the cylinder. Several factory ignition testers eliminate the hazard by confining the spark behind a transparent window.

Occasionally one encounters an engine with a misfire that no amount of carburetor work will cure. In these cases, it can be helpful to monitor spark output with an automotive timing light. The new Xenon lights with a self-contained power source are ideal for small-engine work; otherwise, the light must be powered. Power the light with a 12V auto battery. If you run across an old-fashioned neon timing light in a garage sale, buy it. These lights seem to last forever and require no external power source.

Once you have determined that the ignition is at fault, isolate the ignition module from other wiring. At the minimum, a wire from the primary side of the ignition module goes to kill switch. Verify that the switch is normally open and not grounded to the engine. You can do this with an ohmmeter, or simply by testing for spark with the switch disconnected from the ignition module.

Flywheel

If there is no or weak spark, remove the flywheel in order to check the condition of the key, which fixes the timing between flywheel magnets and the ignition coil. Keys shear or deform in response to sudden loads or as the result of insufficient torque on the flywheel hold-down nut (Fig. 3-1).

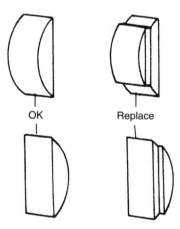

OK   Replace

FIG. 3-1. *A distorted or sheared flywheel key is a major cause of ignition problems.*

The flywheel secures to the crankshaft with a taper, a nut and lock washer. The nut has a conventional right-hand thread 99.9% of the time. Unless you have access to an impact wrench, the flywheel must be prevented from moving as the nut is loosened. Hold the wheel with a strap wrench or lock the piston with a length of nylon rope fed into the cylinder through the spark-plug port. (Fig. 3-2)

FIG. 3-2. *The old rope trick.* Robert Shelby

The flywheel and crankshaft interfaces are tapered, which means that an ordinary gear puller cannot be used. Many flywheels are drilled and tapped for a hub puller as shown in Fig. 3-3. A steering-wheel puller, available from auto parts stores, will usually substitute for the factory tool. You will, however, need metric bolts (usually 6 mm × 1.0) for attachment to the hub.

Flywheels for small motorcycles often have internal hub threads for which suitable pullers are difficult to find. If the dealer is of no help, try looking for a suitable tool in upscale bicycle shops. Otherwise, a metric-threaded puller will have to be machined.

Flywheels with no provision for puller engagement must be shocked off. The preferred tool is a fairly heavy steel bar with threads that match those on crankshaft. Knockers can, at least for now, be purchased from Tecumseh dealers as PN 570105 for right-handed 1/2-in. × 20 threads and PN 670169 for 7/16-in shafts. Metric knockers must be fabricated, which can be done easily on a lathe or more laboriously with a portable drill and the appropriate tap.

Figure 3-4 illustrates the use of a knocker. Wear eye protection, run the knocker down to within two or three threads of the flywheel, and strike hard

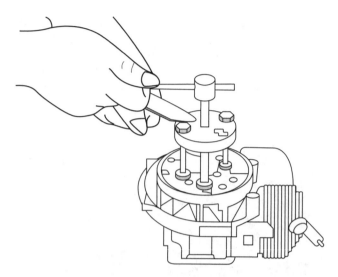

FIG. 3-3. *Tanaka and other professional-quality products have their flywheel drilled and tapped for hub pullers.* Hitachi Koki USA

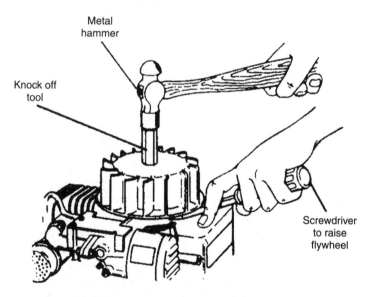

Metal
hammer

Knock off
tool

Screwdriver
to raise
flywheel

FIG. 3-4. *Knocking a flywheel off can result in a broken crankshaft, brinelled main bearings, and scrambled flywheel magnets. But millions of engines have been abused in this manner without such catastrophic effects.*

enough to compress the crankshaft taper. Prying up on the underside of the wheel with a screwdriver assists removal. Be careful where you place the screwdriver when the ignition coil lives under the flywheel. If removing the flywheel in this manner causes the crankshaft to bind, restore main-bearing float by driving the crank upward with a soft mallet (Fig. 3-5).

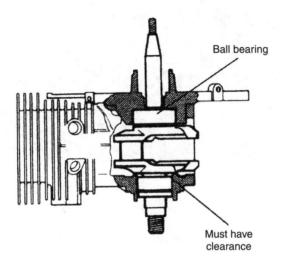

Ball bearing

Must have
clearance

FIG. 3-5. *If the crankshaft binds after shocking the flywheel loose, the main bearings have been unseated. Restore bearing float with a sharp blow from a mallet on the pto (power takeoff) end of the crankshaft.*

An even more brutal approach is to knock the flywheel off with a hammer and brass bar. Back the flywheel nut out to shield the threads.

Most engine makers reference the flywheel to the crank with a removable key. Poulan and others cast the key as part of the aluminum hub (Fig. 3-6). If the key shears, the easiest recourse is to purchase a new flywheel. But flywheels are expensive, and you may want to file a keyway into the hub, using the mark left by the sheared key as the reference. Cutting an internal keyway by hand sounds primitive, but was standard practice throughout most of the nineteenth century. Hardware stores carry key stock that can be filed to fit.

Another, and less reliable, approach is to forget about the key, which is merely an assembly reference. Coat the crankshaft taper with red Loctite, align the keyways, and tighten the hold-down nut.

Warning: Always replace a cracked or otherwise damaged flywheel. Cracks, most of which originate at the keyway, grow slowly until they reach critical length. When that happens, the flywheel explodes.

Assemble the flywheel and crankshaft tapers dry, since the effectiveness of the connection depends upon friction. Although small-engine manufacturers

FIG. 3-6. *Flywheels with stripped cast-in keys can be salvaged with a suitable file and patience.* Robert Shelby

are coy about this, automotive practice calls for torqued threads to be lubricated with 30W motor oil. Broadly speaking, the flywheel nut on mini-motors, those that displace between 22 and 50 cc, requires about 175 in./lb of torque. A reasonable figure for engines in the 80 to 100-cc class is 275 in./lb. Contact your dealer for the exact specification.

Warning: Do not tighten flywheel nuts with an impact wrench. Overtorqueing can crack the hub.

E-gap

All engines have provision for adjusting the E-gap, or the distance ignition-coil armature stands off from flywheel rim. The narrower the gap, the stronger the magnetic field and the greater the voltage produced by the coil. Ideally, the gap should approach zero, but we must allow for bearing play, thermal expansion, and flywheel out-of-round. An E-gap of 0.012 to 0.014 in. is sufficient. Skid marks on the flywheel rim from contact with the coil armature mean that the gap is too narrow or that the crankshaft is bent.

The adjustment procedure goes as follows:

1. Rotate the flywheel to bring the magnets adjacent to the armature.
2. Insert the appropriate gauge between the flywheel and coil armature. A nonmagnetic feeler gauge would be the best choice, although many mechanics use a business card (Fig. 3-7).

FIG. 3-7. *An external ignition coil simplifies E-gap adjustment. Nearly all under-flywheel coils have an adjustment provision by way of a window in the flywheel. Those few that don't can be set by temporarily applying a single layer of friction tape over the ends of the coil armature. The correct E-gap generates a slight, but perceptible, drag on the tape as the flywheel is turned.*
Robert Shelby

3. Loosen the coil hold-down screws. The magnets will draw the coil into specification.
4. Tighten the screws and rotate the gauge free.

Magnetos

Ignition and generating systems depend upon the interplay between magnetic lines of force, or flux, and electricity. This interplay can be described as follows:

- Electricity is generated in a conductor when magnetic lines of force move through a conductor. The faster the movement, the greater the electrical potential generated. To make things more efficient, the conductor consists of a coil of copper wire wrapped over a laminated iron core. Iron acts as a lens to focus magnetic flux.
- The reverse also holds, that is, current flow generates magnetic lines of force at right angles to the conductor. An ignition coil consists of two windings. The primary winding, made up of several hundred turns of heavy wire, goes on next to the armature. The secondary winding, consisting of 10,000 or more turns of superfine wire, wraps

over the primary windings, so that it is saturated with the magnetic lines of force created by current flow in the primary.
- The ratio of primary to secondary windings determines the voltage gain. Thus, if the primary consists of 200 turns and the secondary 10,000 turns, the voltage in the secondary would theoretically increase (10,000/200) 500 times. In practice, the gain is somewhat less.

A magneto generates voltage in the primary by virtue of the moving fly-wheel magnet. One end of the winding grounds to the engine frame; the other end connects to a switch that, when closed, also grounds to the engine. While the classic magneto used contact points as the switch, more modern designs employ a solid-state switch. When the switch closes, the primary circuit is completed through engine grounds and current flows.

The grounding provision for the secondary is similar, in that one end of the winding permanently grounds to the engine and the other end terminates at the spark plug. Once sufficient voltage is present, the secondary circuit completes itself to ground by discharging across the spark-plug gap.

In order to induce the high voltage necessary for ignition, magnetic lines of force must move far more rapidly than afforded by the velocity of the flywheel rim. How a magneto achieves this is a tribute to human ingenuity. Two mechanisms are involved: flux reversal and flux collapse.

Flux reversal accounts for the distinctive E-shape of the magneto armatures. Lines of force flow from the magnetic north pole to the south pole. As you can see in Fig. 3-8, the initial movement is up through the center leg of the armature and down to the south pole through the trailing leg.

As the flywheel continues to turn, the north pole comes into proximity with the leading leg of the armature. Seeking the south pole, magnetic lines of force now move downward in the center leg (Fig. 3-9). Movement is reversed, which means that lines of force have momentarily halted and accelerated in the opposite direction.

At the peak of magnetic activity caused by flux reversal, the switch that controls the primary circuit opens. Current, now denied ground, no longer flows in the primary windings. The magnetic field that accompanied current flow collapses in on itself at light speed. Secondary voltage increases until the spark plug resolves the tension by firing.

Magnetos with contact points

Vintage magnetos employed a mechanical switch, called the contact points, point set, or breaker points, to turn primary current on and off. Some Honda four-strokes still use this archaic form of ignition.

One of the contacts is fixed and permanently grounded to the engine. The moveable contact forms part of the primary circuit. When the points are closed, the primary circuit completes to ground. At about 20° of crankshaft rotation

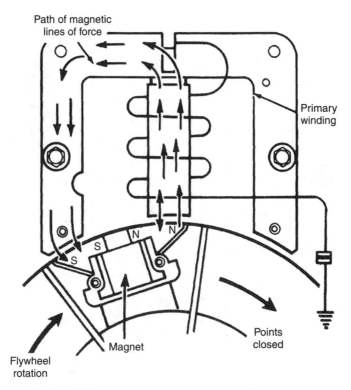

Path of magnetic
lines of force

Primary
winding

N

N

S

S

Points
closed

Magnet

Flywheel
rotation

FIG. 3-8. *The center leg is a two-way street for lines of magnetic force. Initially movement is up through the center leg and down through the trailing leg. Points are closed at this time.*

before top dead center (TDC), the moveable contact cams open, or breaks, to initiate ignition. A condenser (the obsolete term for "capacitor") absorbs electrical energy that would otherwise appear as sparking across the point contacts.

Mechanical switches, especially switches that cycle once every crankshaft revolution, are inherently unreliable. The rubbing block that bears against the cam wears, throwing the points out of adjustment, and the tungsten contacts oxidize and pit. In order to provide a reasonable life, primary current must be limited to something less than 3 A. This limitation results in a low-energy spark, especially during cranking.

Magneto service As with any ignition system, begin by changing out the spark plug and, if that doesn't help, lift the flywheel to inspect the key. Replace points and condenser as a matter of course. All small-engine magnetos that the writer is aware of function with the point gap set at 0.020 in., although some magnetos seem to perform better at 0.017 or 0.018 in. Follow this procedure:

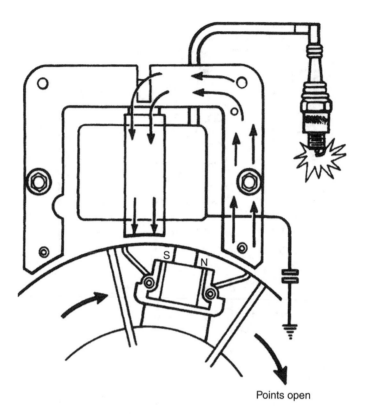

Points open

FIG. 3-9. *A few degrees of additional flywheel movement cause a sudden reversal in flux direction through the center leg. The points open and the spark plug fires.*

1. Disconnect the condenser and point set from the primary winding. Note the lay of the parts.
2. A small screw and one or more locating pins secure the point set to the magneto stator plate. If two screws are present, one is an eccentric that varies the gap between the moveable and fixed contacts.
3. Remove all traces of oil from the stator to provide a clean ground for the new parts.
4. Install the new point set and condenser. Owners of vintage engines may find that these sacrificial parts are in short supply. Any point set that can be modified to fit and any ignition condenser will work.
5. Turn the crankshaft so that the point cams fully open (Fig. 3-10). The hold-down screw will have to be lightly snugged down to prevent the point spring from closing the points. Now bring the moveable point to within 0.020 in. of the fixed point. Breaker sets without an adjustment screw are slotted for screwdriver access.

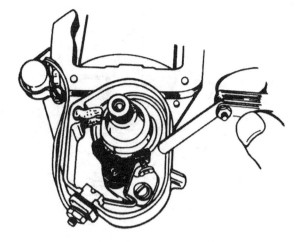

FIG. 3-10. *Wipe any oil off the feeler gauge and, if you follow the procedure favored by many mechanics, adjust the gap by bracketing. A 0.020-in. gap is indicated when a 0.019-in. blade slips easily between the contacts and a 0.021-in. blade generates perceptible drag. The blade must, of course, be held parallel to the contacts.*

6. Test for spark. In event of no spark, burnish the contacts with a business card and retest. Tungsten contacts oxidize in storage and have zero tolerance for oil, even the oil left by a fingerprint.
7. If a new condenser and new, properly adjusted, and clean points fail to restore spark, replace the coil.

Timing Magnetos for light motorcycles, outboard motors, and some utility engines have provision for timing adjustment (Fig. 3-11).

Transistorized ignition

Much to the relief of mechanics everywhere, transistorized ignition (TSI) has become nearly universal. Transistorized magnetos substitute a Darlington transistor (actually a pair of transistors) for the contact points and condenser. Doing away with these troublesome components was the best thing that happened to small engines since the advent of loop scavenging.

On the downside, these TSI systems are nonrepairable. If something goes wrong, replacement is the only option. Nor can TSI be easily adapted to earlier engines without the availability of factory parts. The trigger coil requires a dedicated flywheel magnet.

Capacitive discharge ignition

As the twentieth century drew to a close, transistorized magnetos gave way to capacitive discharge ignition (CDI) systems. As the name suggests, CDIs

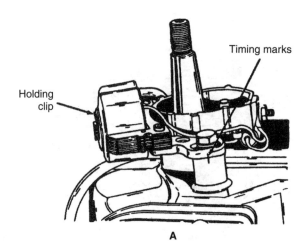

Timing marks

Holding
clip

A

FIG. 3-11. *Phelon magnetos of the type found on American utility engines have timing marks as an assembly convenience (A). High-performance engines should be timed with a dial indicator, such as the Tecumseh PN 670241 (B). The piston is brought up to top dead center, the indicator zeroed, and the flywheel turned until the piston (which should be free of carbon deposits) is at the specified distance before TDC. Timing is correct when turning the magneto stator against the direction of engine rotation just cracks the points open. An ohmmeter with one test lead clamped to the moveable contact arm and the other grounded gives an accurate indication of point break. A piece of cigarette paper inserted between the closed contacts can also be used to detect point separation.*

store electrical energy in a capacitor, which then discharges into the primary windings of an ignition transformer.

Figure 3-12 illustrates the basic circuit. A flywheel magnet generates several hundred volts in the charge coil that is then rectified (i.e., converted from AC to DC) by diodes D1 and D2, and stored in the capacitor (C1). The silicon-controlled rectifier (SCR) initially acts as a check valve to permit current flow into the capacitor and block flow out of it. Further rotation of the flywheel causes the trigger coil to generate a signal voltage across resistor (R1). This voltage sends the SCR conductive. Seeking ground, electrons stored in the capacitor flow into the primary side of the ignition coil, or as its commonly termed the pulse transformer. The expanding magnetic field that accompanies the influx of current induces a sudden spurt of high voltage in the transformer secondary winding. This voltage goes to ground through the spark plug.

The voltage rise in the secondary windings of point-controlled magnetos averages about 50 V/μs (millionths of a second). Transistorized magnetos do a bit better at 80 V/μs. CDIs build voltage almost instantaneously at rates of between 3 kV and 10 kV/μs. Confronted with these sudden bursts of voltage,

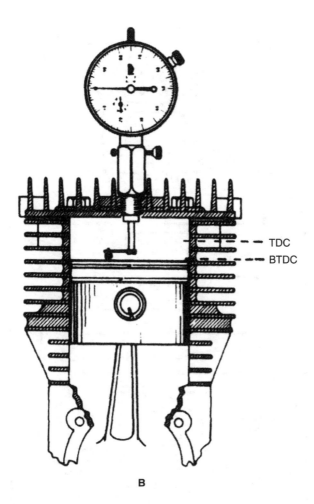

B

FIG. 3-11. *(Continued)*

dirty spark plugs fire before they have time to leak to ground. CDI makes starting easier, especially for two-stroke engines.

On the debit side, CDI voltage disappears almost as quickly as it is generated. Burn time—spark duration—ranges from 50 to 80 μs. In comparison, transistorized ignitions have burn times of between 500 μs and 0.001 second. Consequently, a CDI can fail to reliably ignite lean mixtures. MSD, Mallory, and other high-performance automotive CDIs minimize the problem by generating multiple sparks at speeds below 3000 rpm. Small-engine units do not, at present, have this capability.

TSI and CDI systems require resistance-type spark plugs, if your neighbors are going to talk on cell phones or watch television.

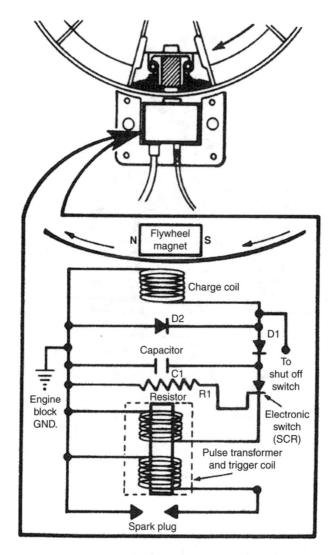

FIG. 3-12. *Tecumseh CDI circuitry is typical, although some battery-operated systems include diodes to protect components from reversed polarity.*

Digital CDI

Solid-state systems have some built-in ignition advance by virtue of flywheel velocity. As rpm increases, signal voltage occurs earlier, since the flux field moves faster. But advance is linear and not tailored to engine characteristics. Digital control retards ignition during cranking to prevent starter-cord "bite"

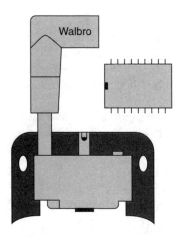

FIG. 3-13. *Walbro digital CDI.*

and advances timing in a patterned sequence that can be modified by the engine maker (Fig. 3-13). The integrated circuit (IC) also acts as governor, retarding timing as the preset rpm limit is approached. Digital control costs almost nothing at the OEM level: one company offers IC timing controllers for 39 cents each in batches of 10,000.

But like other solid-state components, digital CDIs are nonrepairable and nontestable. If you suspect that something is wrong, say, that the timing has gone haywire, the only recourse is to spring for a new module.

Spark plugs

Now that leaded fuel is no more, spark plugs can be cleaned by soaking in carburetor cleaner. Sandblasting is out of the question, since the abrasive particles imbed themselves in the insulator where they fall into the cylinder once the engine starts.

In passing, it should be mentioned that the jagged, blackish discoloration on the outside of the insulator just below the rubber boot has no effect upon performance. It is caused by the same corona effect that makes high-voltage power lines glow.

The center electrode rounds off in service, increasing the voltage requirement. Ordinary carbon steel or copper electrodes can be dressed with a point file. This does not apply to exotic plugs sometimes used in high-output engines. These plugs are characterized by small-diameter center electrodes that, in order to provide useful life, are capped with precious metals. Filing merely removes the gold, platinum, silver, or yttrium coating. Champion iridium electrodes can be filed, since the whole electrode is made of this refractory material.

Traditional spark plugs are best gapped with a cam-type gauge as illustrated back in Fig. 2-2. Precious-metal plugs generally come pregapped, but can be set to the exact specification if the vulnerable center electrode does not see any bending force. In other words, confine the bending to the side electrode by using long-nosed pliers or a small bending bar, drilled to slip over the side electrode.

Remove all traces of grease and oil from the gasket surface (aerosol brake cleaner and Q-tips work) and run the spark plug in by hand for at least three full turns. Torque to specification

10 mm	7.2–8.7 ft/lb
12 mm	10.8–14.5 ft/lb
14 mm	18.0–21.6 ft/lb

While some mechanics like to use antiseize compound on outboards and other engines exposed to weather, these lubricants reduce friction enough to result in over-torqueing. Silicon, because of its poor lubricating characteristics, is a good compromise.

Heat range

Heat range depends primarily upon the exposure of the insulator to combustion heat. As shown in Fig. 3-14, the greater the exposure and the longer the thermal path, the hotter the plug. The plug should run hot enough to prevent fouling, but not so hot that it ignites the mixture early, before the spark (Fig. 3-15). Should this happen, the engine self-destructs.

Manufacturers describe heat ranges numerically. For domestic spark plugs, the higher the number, the hotter the plug. Thus, a Champion J19LM is hotter than a J17LM. Foreign manufacturers usually, but not always, assign higher numbers to colder plugs. At the first sign of trouble, replace the spark plug with the same type and, whenever possible, with the same brand specified in the owner's manual. Interchange guides list physically equivalent plugs, but heat ranges vary somewhat between manufacturers.

Matching spark plugs to the needs of racing and other high performance engines is beyond the scope of this book. Plenty of this information is available elsewhere. But more prosaic engines sometimes come without documentation and one needs to select a spark plug that will do no damage. Begin with a cold plug and run the engine under normal work loads for an hour or so, adjusting the carburetor for best power. If the plug blackens and fouls, go hotter one heat range. Continue a step at time with spark plugs from the same manufacturer until the insulator takes on a dark brown color, indicating that temperatures are normal. Err on the side of cold (dark), since the worst that a cold plug can do is foul.

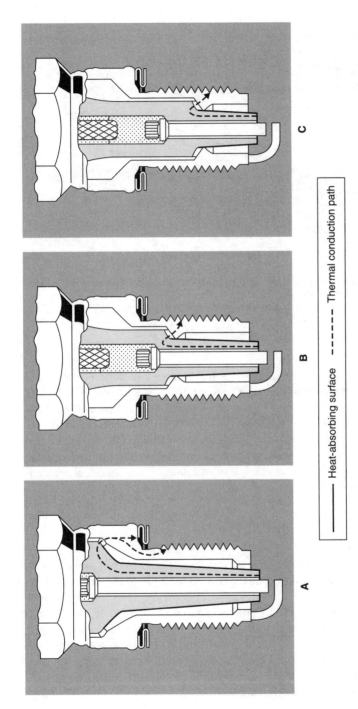

FIG. 3-14. *The shape of the insulator tells the story. A hot plug exposes a large surface of the insulator to combustion heat that must travel high into the barrel to find release (A). The insulator for a cooler plug has less exposure and provides a shorter thermal path (B). The coldest plug has the smallest insulator nose (C). Bosch*

—— Heat-absorbing surface   - - - - - Thermal conduction path

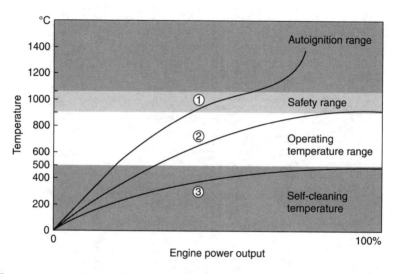

FIG. 3-15. *A spark plug with the correct heat range (2) walks the narrow line between fouling and autoignition.* Bosch

Thread repair

Over-torqueing or cross-threading strips the threads on aluminum cylinder heads. With the right tools, one can salvage the head with a Heli-Coil or equivalent insert. Unfortunately, the cost of the installation tools is prohibitive for weekend mechanics, who may see a stripped head once in a lifetime. Spark-plug thread repair kits are specific to each thread diameter and cost $130 or more. Small-engine dealers invoke a daunting minimum labor charge for this 5-minute job. The best chance of obtaining an economical repair is at an old-fashioned auto parts house that does machine work on the side.

Summary

A new spark plug cures most ignition problems. Other likely suspects are a damaged flywheel key or fretted insulation on the wire to the kill switch. As a last resort, replace the solid-state ignition module. The tungsten points used in vintage magnetos are an endless source of trouble. The point gap narrows as the rubbing block wears and the contacts become increasingly resistive in service. "New old-stock" point sets, which have languished on dealer shelves for decades, can be restored to service by burnishing the contacts with a business card. If you cannot find the correct replacement condenser, substitute an automotive unit.

4

Fuel systems

Something like 90% of engine malfunctions that a new spark plug can't fix are caused by the fuel system. The system consists of the fuel tank, filter, pulse and fuel lines, air filter, carburetor, reed valve (when fitted), and engine crankcase. We will discuss each of these components separately.

Fuel tank

Routinely examine the tank for leaks and for loose or stripped hold-down bolts. Mysterious engine shutdowns that elude obvious explanation may be caused by a failed tank vent, which can be integral with the fuel cap or, as in the case of the Tanaka shown in Fig. 4-1, a separate component.

Fuel filters

Fuel filters should be changed at least every 100 hours of operation and whenever

- Fuel delivery, signaled by no fuel to the carburetor or by a bleached spark-plug, has been compromised,
- Fuel quality is suspect.

Use a hooked wire to retrieve in-tank filters of the kind used on hand-held tools. Replace deformed or rusted hose clamps.

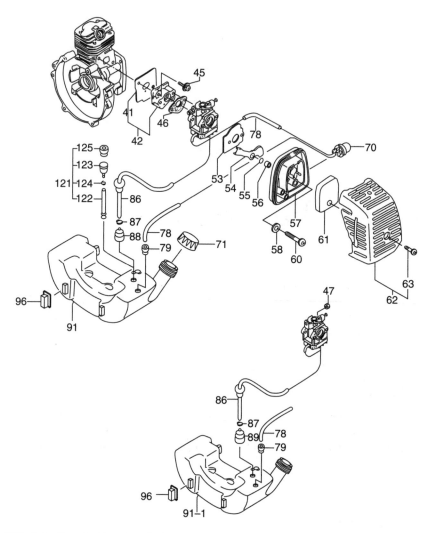

FIG. 4-1. *Tanaka TBC-250 fuel-tank assembly exhibits the sort of detailed engineering found in professional equipment. Spring clips secure the fuel filter (89) and the check valve (123). Fuel pipes (78 and 86) make up to a flexible grommet and, for added leak protection, draws from the top of the tank.*

Fuel lines

Larger two-strokes use elastomer hose that, in its latest evolution, has a nylon core to reduce evaporative fuel losses. These hoses come in 1/2-, 5/16-, and 3/8-in. IDs.

Be skeptical of the plastic lines used on portable equipment to transfer fuel and energize the fuel pump with crankcase pressure pulses. Heat and vibration harden the plastic, which then leaks fuel and, what is more difficult to diagnose, air. Pulse lines are also susceptible to clogging.

Most applications call for 0.110-in. ID by 0.190-in. OD or 0.140-in. ID by 0.230-in. OD lines. To avoid confusion, purchase replacements from a dealer for the engine in question.

The plumbing can be complex. Make a sketch or take digital photos of the routing, which should be well clear of the muffler and other hot spots. Confusing primer or pulse lines with fuel lines prevents the engine from starting.

To make line-to-tank installation easier, cut end of the plastic line at an angle and lubricate it with motor oil. You may want to stiffen the line with wire. Once the line has passed through the tank, remove the gas cap and use long-nosed pliers or a surgical hemostat to pull several inches out of the tank (Fig. 4-2). Trim the end square and install the filter together with the optional hose clamp. Check for leaks before starting the engine and after 30 minutes or so of hard use.

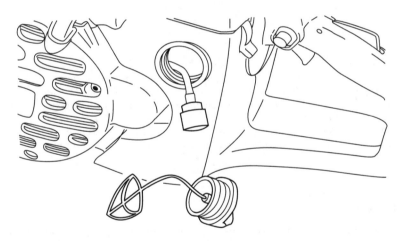

FIG. 4-2. *As this Wacker Neuson illustration shows, there should be enough slack in the line to allow the filter freedom to respond to gravity as the tool is tilted off the horizontal.*

Air filters

There is some dispute about which filters are better—pleated paper or urethane foam. In theory, paper catches smaller particles, but some field tests indicate that foam reduces engine wear. Foam filters consist of a labyrinth of tiny passages that trap solids along their whole depth. Paper filters, on the other hand, present a surface barrier, which can quickly clog.

Light-duty two-strokes make do with a single foam filter, while better engines combine both types. A foam prefilter stops large particles before they impact the paper main filter. STIHL cutoff saws combine centrifugal filtering with two stages of conventional filtering. Husqvarna uses a similar arrangement on its concrete saws. The company claims that the centrifuge removes 80 to 90% of the dust these tools generate prior to filtration. When the filters clog, Husky carburetors go lean to compensate for the increased pressure drop.

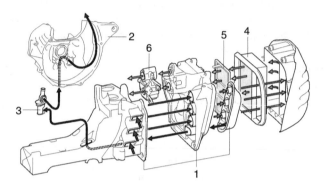

FIG. 4-3. *The STIHL TS 410 cutoff tool combines centrifugal with conventional filtering. The tank housing (1) acts as a centrifuge to separate out larger and heavier dust particles. Contaminated air passes through the molded hose (3) to the fan side of the crankcase (2) and into the atmosphere. Pre-cleaned air undergoes two stages of filtering (4) and (5) before entering the carburetor (6). This system requires very little maintenance. Change the filter only if the engine shows loss of power or if more than a year of extended used has passed since the last filter change. Attempts to clean the filter damage it.*

Inspection

As filters collect dirt, they become more restrictive. The increased pressure drop sends the carburetor rich. The engine loses power, the spark plug turns darker, and the exhaust becomes acrid. Paper filters are also sensitive to moisture. Some filter housings give little protection against rain and any can be defeated with a pressure washer. When this happens, the cellulose fibers swell and close off the air channels. Solvents or oil leached from the prefilter have a similar effect. When in doubt, replace the paper element.

Servicing

Remove the filter cover and carefully wipe off dust on the housing with a moist shop towel. Dust entry during filter service is a major cause of engine damage. Wash foam filters in warm water and detergent. Rinse, dry, and apply the proper wetting agent, available from Husky and other dealers. Motor oil quickly gets sucked into the carburetor.

Paper filters are throwaways, impossible to clean. Blowing out the filter with compressed air releases a satisfying cloud of dust, while at the same time blasting microscopic holes in the media. Replace the filter with a factory (and not a Chinese aftermarket) part.

Carburetors

In a chemically perfect world, engines would consume 14.7 grams of air for each gram of fuel. This is called the stoichiometric ratio. Unfortunately, the ideal can only be approximated a part throttle and under light loads. For best power, a two-cycle engine needs a mixture ratio of between 12 and 13:1. And, if the engine is to live at wide open throttle, the mixture should be richened further to 11 or even 10:1. The surplus gasoline functions as a coolant. Cold starting and idle require even richer mixtures as indicated in Table 4-1.

Table 4-1
Air-fuel mixture ratios for different
operating conditions

Cold start	1:1–4
Idle	1:8–10
Low speed	1:10–13
Cruise, light load	1:14–15
High speed, heavy load	1:10–13

To accommodate the wide range of mixture requirements, carburetors employ at least two fuel circuits: high-speed and low-speed. In addition, carburetors have a restriction, called a *venturi*, to build vacuum in the throttle bore and a throttle valve to control engine speed. A diaphragm or float valve regulates the level of fuel in the instrument.

Diaphragm carburetors

Because diaphragm carburetors are insensitive to attitude—the diaphragm functions in any position, even inverted—they are standard on trimmers, hedge clippers, and other handheld tools. The trade-off is less-than-perfect control over the mixture and reduced reliability.

Sometimes a carburetor can be returned to service by merely replacing the diaphragms and squirting carburetor cleaner into places that look

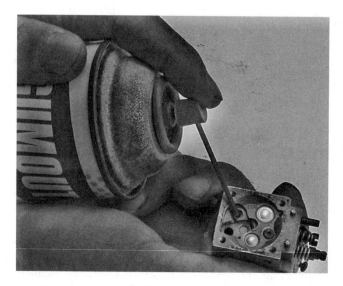

FIG. 4-4. *New diaphragms and carburetor cleaner can sometimes "fix" a carbure-tor to the extent that the engine starts and runs.* Robert Shelby

important (Fig. 4-4). But if repairs are to be more than a matter of luck, one needs to understand what goes on in these little cubes of aluminum.

Critical areas are the diaphragms, inlet needle, low- and high-speed circuitry, and the provision for cold starting. All carburetors have these elements, which work in the same general fashion. Once you have come to terms with them, you should be able to repair Tanaka, Tecumseh, Zama, and all but a handful of Walbro carburetors.

What remains are bells and whistles, such as the accelerator pumps and speed-limiting governors found on high-end instruments. While these features have technical interest, they exert little impact upon repair work.

Fuel pump   Most handheld engines employ a fuel pump, mounted on the carburetor body (Fig. 4-5). The pump element consists of a nitrile diaphragm, with flaps cut into it that act as inlet and output check valves (Fig. 4-6). The cavity cast into the pump body is a variable-displacement fuel reservoir, whose volume changes as the diaphragm, which forms the floor of the cavity, flexes.

As the piston rises toward top dead center (TDC), it leaves a partial vacuum behind it in the crankcase (Fig. 4-7). This vacuum, transmitted to the outboard side of the diaphragm by a drilled passage in the carburetor body or by an external line, pulls the diaphragm down, away from the fuel cavity. Because of the increase in its volume, a slight vacuum is created in the cavity. Fuel, under atmospheric pressure in the tank, rushes in past the open inlet valve to fill the void. The pump is now charged.

FIG. 4-5. *Pumps on smaller carburetors usually are on the side opposite the metering diaphragm.* Robert Shelby

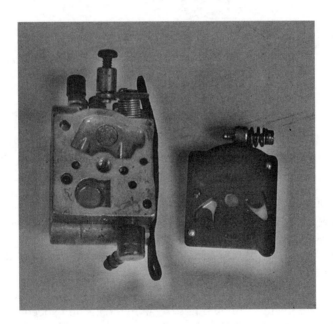

FIG. 4-6. *Pump with cover removed. The flap-like cutouts in the diaphragm function as inlet and outlet valves.* Robert Shelby

A few milliseconds later, the piston rounds TDC. As it travels down the bore, it pressurizes the crankcase to 5 or 6 psi above atmospheric. The pressure forces the pump diaphragm upward into the cavity (Fig. 4-8). The inlet check valve closes to prevent fuel from cycling back to the tank. At the same time, the discharge valve opens, admitting pressurized fuel into the carburetor.

Fuel pump intake

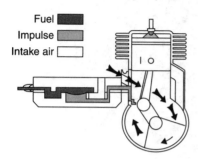

Fuel
Impulse
Intake air

FIG. 4-7. *As the piston rises, it evacuates the crankcase. The negative pressure pulse, acting on the underside of the diaphragm, pulls the diaphragm down and away from the fuel cavity. During this phase, the inlet flap valve opens and the outlet valve closes. Impelled by atmospheric pressure, fuel enters the cavity.* Walbro Corp.

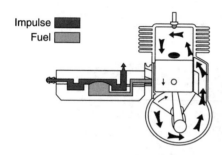

Impulse
Fuel

FIG. 4-8. *Crankcase pressure forces the diaphragm upward, opens the outlet valve, and closes the inlet valve. Fuel flows into the carburetor body.* Walbro Corp.

Pumps fail progressively, pumping less fuel as the diaphragm hardens and loses elasticity. Loss of power under full load is the first symptom. However, a partially clogged main jet or inlet screen (including the one sometimes found downstream of pump discharge) can have the same effect. Air leaks at fuel- or pulse-line connections usually are severe enough to prevent starting. The same holds for obstructions in the pulse line. Another cause of failure is incorrect assembly: the diaphragm and gasket must be installed in the correct sequence. Some pumps go together with the gasket against the carburetor body, followed by the diaphragm. Others reverse the sequence.

Metering   Metering is the term used to describe the ability of the carburetor to regulate mixture strength under varying loads and speeds. The position of the throttle determines the volume of air available for combustion; four components control the amount of fuel supplied to the jets (Fig 4-9):

- Metering diaphragm
- Inlet needle
- Inlet lever
- Lever spring

Unlike the fuel pump, which dances to crankcase pulses, the metering diaphragm responds to variations in inlet-pipe vacuum. The underside of

FIG. 4-9. *Most carburetors transfer movement from the diaphragm to the needle by way of a spring-loaded lever. The lever is the bright metal part shown at 2 o'clock in the photo.* Robert Shelby

the diaphragm vents to the atmosphere; the upper, or wet, side sees the same vacuum that draws fuel into the carburetor bore. During the moments of peak vacuum the diaphragm lifts and presses against the lever. This action overcomes the spring tension that holds the needle closed. The needle then unseats in response to 5 or 6 psi of pressure created by the fuel pump. Fuel flows past the unseated needle into the space above the diaphragm, known as the metering chamber.

A few milliseconds later, the inlet port closes. Manifold vacuum drops, causing the diaphragm to relax its pressure on the lever. The inlet needle, now under spring tension, closes to cut off fuel delivery.

Fuel delivery depends upon two variables: the tension of the lever spring and lever height. Substituting a weaker spring for the stock item increases fuel flow by reducing the workload on the diaphragm. A similar effect can be had by raising the lever, so that it comes into play earlier and allows the needle to stay open longer.

In addition to supplying fuel for steady-state operation, the metering system must provide the rich mixtures needed for acceleration and inhibit fuel delivery during coastdown. How well the engine behaves during these transitional modes depends upon the resilience of the diaphragm, the relationship between the lever and diaphragm, and spring tension. While factory defaults are not written in stone, one should exert extreme caution when swapping springs and experimenting with different lever heights.

But not all of these instruments employ a lever. Figure 4-10 shows the inlet needle and associated parts for a Tecumseh carburetor. The diaphragm bears

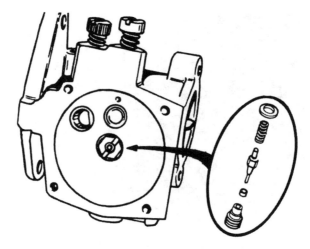

FIG. 4-10. *Tecumseh/Tillotson direct-acting inlet needle.*

directly against the needle without the intermediary of a lever. The Tillotson also has a replaceable needle seat, a welcome feature not always found on these carburetors.

The relationship between the diaphragm gasket and diaphragm is critical. Some carburetors assemble with the gasket next to the carburetor body; others have the gasket next to the cover. It pays to have a notebook handy when disassembling unfamiliar equipment.

A hole in the metering-chamber cover vents the underside of the diaphragm to the atmosphere. The vent also enables the carburetor to be primed for easier starting. This is done by inserting a small Allen wrench into the vent and gently pressing the diaphragm upward to unseat the inlet needle.

The inlet needle causes a disproportionate amount of trouble. A tiny spring, which looks like it could have come from a watch, holds the needle closed. To generate adequate unit pressure, the needle contacts the seat along a narrow band. A microscopic spec of dirt is enough to compromise the seal. And because all that opens the needle is fuel-pump pressure acting against its tip, the needle tends to stick closed. The good news is that a squirt of aerosol carburetor cleaner is usually enough to set things right.

Venturi   A carburetor bore resembles an hour glass, wide at the ends and narrow at the center. The necked-down section, known as the venturi, causes the incoming air stream to accelerate and lose pressure (Fig. 4-11). To understand why, consider that the venturi is open at both ends. As much air leaves as enters, but the restriction increases the distance air travels, which means that air velocity must also increase. Since nothing in nature is free, velocity comes at the cost of pressure.

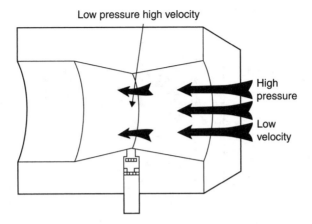

Low pressure high velocity

High pressure

Low velocity

FIG. 4-11. *A venturi generates negative pressure and increased velocity.* Walbro Corp.

The same principle explains why airplanes fly. The upper surface of the airfoil is curved and the lower surface flat. Air passing over the top of the wing loses pressure relative to air moving under it. The pressure differential is felt as lift.

High-speed circuit   The vacuum created by the venturi draws fuel from the metering chamber through the high-speed jet and discharge nozzle (Fig. 4-12). The nozzle, located near the narrowest point of the bore, begins to flow at around quarter throttle. As the throttle opens wider, the vacuum produced by increased air flow draws greater amounts of fuel through the nozzle to keep pace with engine demands.

In the past, nozzles were threaded into the carburetor body and could be removed for servicing. Most modern carburetors have pressed-in nozzles that should not be disturbed.

The size of the main jet determines the rate of fuel delivery through the nozzle. Depending upon the carburetor, the jet may be fixed or adjustable, removable or not. But all main jets have one characteristic in common: as the most restrictive component in the high-speed circuit, they are the first to clog.

Low-speed circuit   On most carburetors, the low-speed circuit discharges through a row of holes drilled into the carburetor body, adjacent to the throttle plate. The nomenclature varies: most technicians refer to the first hole, the one closest to the engine, as the idle port. Those that remain are called progression or transition ports.

At idle, the throttle is almost closed, a condition that divides the carburetor bore into two pressure zones. Low pressure exists on the engine side of the

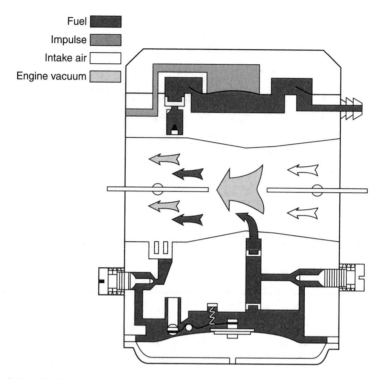

FIG. 4-12. *The high-speed circuit flows in response to venturi vacuum.* Walbro Corp.

throttle and atmospheric pressure on its upstream side. Impelled by the low pressure, fuel enters through the idle port (Fig. 4-13). Air, under atmospheric pressure, enters through the transition ports to mix with the fuel before discharge. Aerating the fuel assists in atomization and helps prevent fuel puddling in the intake pipe and crankcase.

As the throttle cracks open, the transition ports come under inlet-pipe vacuum and pass the additional fuel needed as the engine speeds up (Fig. 4-14). At some point near quarter throttle, air flow through the venturi generates enough vacuum to energize the high-speed circuit.

The idle jet works in tandem with the idle and transition ports to determine the amount of fuel delivery. Most idle jets, even those on emissions carburetors, have some limited provision for adjustment: backing the adjustment needle out of the jet orifice richens the mixture by allowing more fuel to flow for the same amount of air.

Nozzle check valve The discharge nozzle incorporates a one-way valve, in the form of a ball or a plastic disk. At mid- and high speeds, air flow through

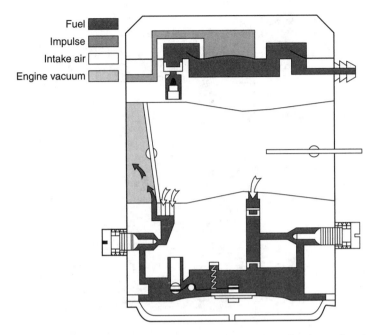

Fuel
Impulse
Intake air
Engine vacuum

FIG. 4-13. *At idle, the throttle plate is nearly closed and fuel flow is confined to the idle port. Transition ports, upstream of the idle port, aerate fuel awaiting discharge in the idle pocket. The low-speed jet shown here is adjustable.* Walbro Corp.

the venturi generates enough vacuum to unseat the check valve and permit fuel to flow through the nozzle (Fig. 4-15).

At idle and part throttle conditions are reversed. The venturi comes under atmospheric pressure and the pressure in the metering chamber drops as the low-speed circuit consumes fuel. The check valve closes. Were the valve to remain open, air under atmospheric pressure would enter the metering chamber via the nozzle. The metering diaphragm, denied the low pressure necessary to unseat the inlet needle, would hang motionless. A sticking nozzle check valve has no effect upon high-speed operation. But the engine quickly runs out of fuel at idle.

Some Walbro carburetors substitute a fine-mesh screen for the check valve. Surface tension of fuel on the mesh is enough to prevent air entry into the nozzle. Should the screen clog, fuel will be denied at wide throttle angles.

Cold start The reluctance of gasoline to vaporize at ambient temperatures coupled with low cranking speeds, makes cold staring difficult. Extremely rich mixtures are required.

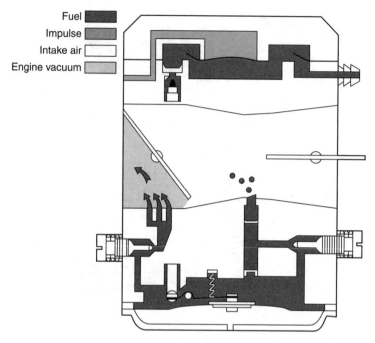

FIG. 4-14. *Off-idle operation draws fuel from the idle port and the two transitional ports. Greater throttle angles energize the venturi and the high-speed circuit. The degree of overlap between low- and high-speed systems varies with carburetor design.* Walbro Corp.

Choke The simplest way to increase fuel flow is to seal off the carburetor bore with a choke valve upstream of the venturi. When the valve is closed, all circuits come under vacuum and pass fuel (Fig. 4-16). Nothing could be simpler and less liable to failure. On the downside, the operator must calculate how much choke a semi-warm engine needs. Insufficient choking starves the engine; overchoking drowns it.

The choke also functions as a diagnostic aid: a warm engine that runs better with the choke engaged ingests air downstream of the carburetor. Expect to find loose carburetor hold-down bolts, a torn flange gasket, or leaking crankshaft seals.

Purge pump The heart of the purge system is a rubber bulb that can be mounted on the carburetor or at some position more convenient for the operator. Like any pump, the bulb works in conjunction with inlet and outlet check valves.

Pressing on the bulb opens the outlet check valve to displace any air in the system back through a line to the tank. Releasing the bulb closes the outlet check valve, isolating the tank, and opens the inlet valve to apply

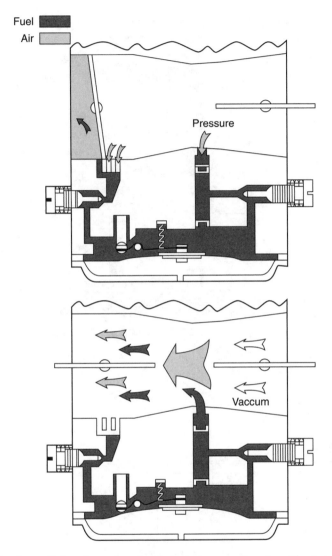

Fuel
Air

Pressure

Vaccum

FIG. 4-15. *At small throttle angles, the venturi comes under atmospheric pressure and the nozzle check valve closes. Opening the throttle increases air flow through the venturi, a condition that generates the negative pressure. When this happens, the check valve lifts and fuel flows through the nozzle.* Walbro Corp.

vacuum to the metering chamber (Fig. 4-17). The metering diaphragm responds to the vacuum, the inlet needle unseats, and fuel flows into the metering chamber, past the open inlet valve, and into the bulb (Fig. 4-18). When the transparent bulb is full of fuel, the carburetor is primed.

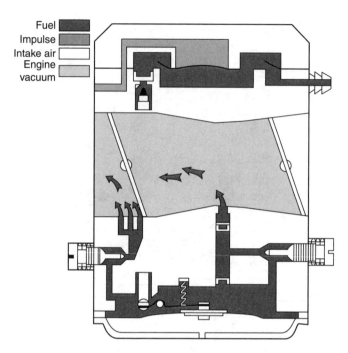

FIG. 4-16. *Cranking with the choke closed evacuates the carburetor bore and all circuits flow.* Walbro Corp.

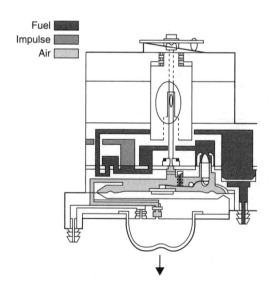

FIG. 4-17. *Releasing the bulb closes the discharge check valve and opens the inlet check valve. The vacuum created as the bulb expands lowers the metering diaphragm and unseats the inlet needle.* Walbro Corp.

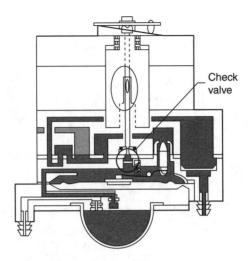

FIG. 4-18. *When the bulb is full of fuel, the carburetor is purged of air and primed to deliver a rich starting mixture. The check valve prevents air from entering the metering system through the discharge nozzle as the bulb inflates. This valve does not replace the nozzle check valve, used on butterfly throttle carburetors as illustrated in Fig. 4-15. An additional check valve blocks air entry from the idle circuit.* Walbro Corp.

Primer pump Purge and primer systems both employ an elastomer bulb as the pump element and cycle entrapped air back to the fuel tank. But, rather than merely lifting the inlet needle, a primer delivers a shot of raw fuel into the carburetor air horn. Over-enthusiastic priming can flood the engine.

Pressing the bulb draws air out of the circuits; as the bulb recovers shape, it fills with fuel. Pressing it again sends fuel into a storage compartment or to a wick. For some carburetors, wick saturation is not automatic; in addition to pressing the bulb twice, the operator must also press the starter button on the underside of the carburetor body (Fig. 4-19).

Accelerator pumps Snap the throttle open and the sudden onrush of air fills the intake pipe. Manifold vacuum momentarily drops to zero. Robbed of vacuum, the carburetor no longer provides sufficient fuel. The inertia of gasoline vapor compounds the problem. The engine hesitates and, in extreme cases, announces its distress with backfires.

The traditional cure for hesitation is to mask the problem with high idle rpm—some two-strokes idle at 3000 rpm—and rich idle mixtures. A better approach, and one more compatible with emission requirements, is to incorporate an accelerator pump.

Piston-type pump Some carburetors employ a spring-loaded brass piston as the pump element. At low speeds, the piston rests on flat spot milled on

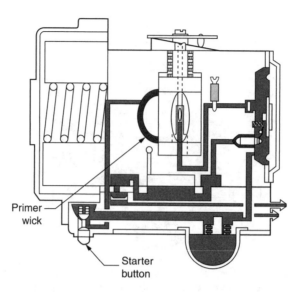

FIG. 4-19. *This Walbro barrel-valve carburetor employs a wick that is wetted by the primer pump when the bulb is pressed twice and the starter button held down.*

the throttle shaft. As the throttle swings open, it cams the piston forward to force fuel into the carburetor bore or metering chamber (Fig. 4-20).

Pulse-driven enrichment Pulse-driven enrichment works by exciting the metering diaphragm with positive pressure pulses from the crankcase. A passage in the carburetor body runs from the crankcase to the throttle shaft. When the throttle opens, a hole in the shaft aligns this passage with a second passage that leads to the dry side of the metering diaphragm (Fig. 4-21). Normally the dry side of the diaphragm sees only ambient pressure. Piping in crankcase pulses increases fuel delivery so long as the throttle remains open.

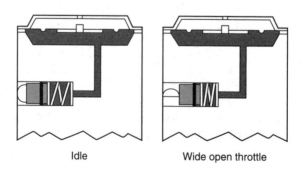

Idle Wide open throttle

FIG. 4-20. *A piston-type pump, keyed to throttle motion, represents the most direct approach to mixture enrichment.* Walbro Corp.

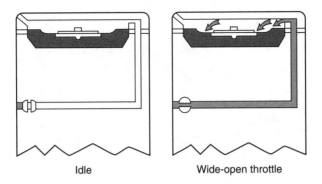

Idle   Wide-open throttle

FIG. 4-21. *Some Walbro carburetors use crankcase pulses to richen the mixture at large throttle angles.*

Boot pump This approach to mixture enrichment uses a flexible boot as the pump element (Fig. 4-22). At small throttle angles, the boot distends with fuel. Opening the throttle exposes the outer surfaces of the boot to crankcase pulses. The pressure differential collapses the boot, forcing fuel into the carburetor air horn.

Governor The governor consists of a spring-loaded ball that vibrates to augment fuel delivery at a pre-set engine rpm. The brass capsule, slotted for a screwdriver, mounts externally some Tillotson and Walbro carburetors.

Most tuners adjust a carburetor by setting the mixture needles halfway between lean and rich drop-off. The high-speed adjustment requires a generous amount of throttle. If a governor is present, it will richen the high-speed mixture as the rpm limit is approached. Tightening the high-speed adjustment needle has little effect upon fuel delivery, making it impossible to find lean drop-off. The way around this problem is to make the adjustment while the engine is under load.

Barrel-valve carburetors Most Zama and some Walbro carburetors regulate fuel delivery mechanically with a barrel throttle and tapered needle that extends into the main nozzle. These instruments dispense with low-speed circuit: all fuel enters through the nozzle.

The Walbro range consists of the WY (Fig. 4-23) and the WZ series (Fig. 4-24), with the former distinguished by the build-in air filter and the right-angle turn entering air makes. Variations of the WY series currently extend in alphabetical order from the WYE through WYP. The WYF is the lone example of a float-type, barrel-valve carburetor. All WYs have pump and metering diaphragms stacked on the same side of the plastic or metal carburetor body.

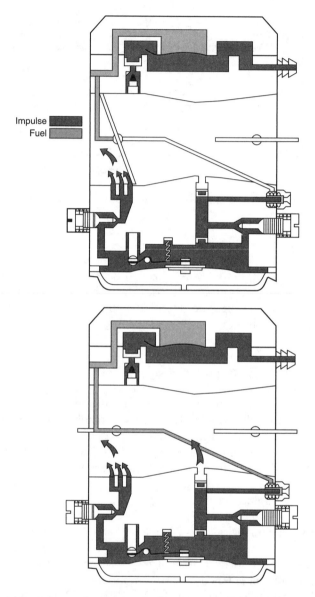

FIG. 4-22. *A boot-type accelerator pump charges at small throttle angles, as shown in the upper drawing. Opening the throttle directs positive crankcase pulses to the dry side of the boot, which empties itself directly into the carburetor bore or by way of the metering chamber.* Walbro Corp.

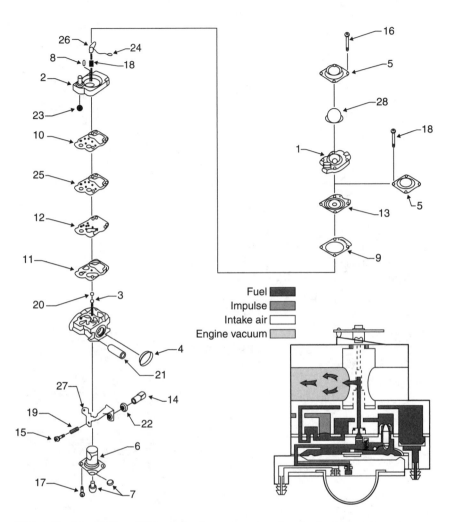

FIG. 4-23. *The Walbro WY barrel-valve carburetor piggybacks the pump and metering chamber. In most examples the jet is fixed and nonreplaceable.*

Figure 4-25 illustrates how fuel and air are metered. The obstruction presented by the barrel creates negative pressure at the discharge nozzle. At low speeds, the barrel (A) and needle (B) restrict air and fuel flow. As the barrel rotates open, it cams upward and lifts the needle higher in the discharge nozzle (C). Retracting the tapered needle increases the effective area of the nozzle and more fuel flows. At higher speeds, the cross-drilled passage in the barrel comes into better alignment with the axis of the carburetor bore. Thus, air delivery keeps pace with fuel delivery.

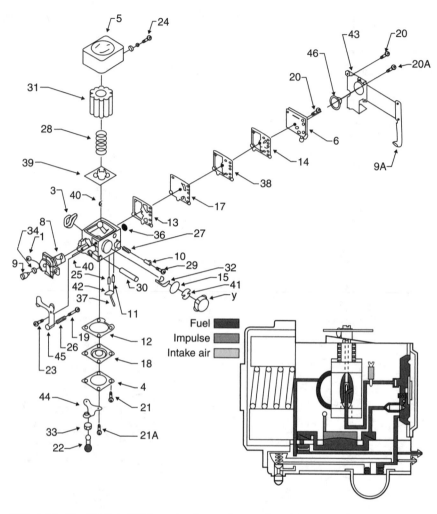

FIG. 4-24. *The Walbro WZ barrel-valve carburetor has fuel pump and metering diaphragms on opposite sides of the main casting. The primer bulb mounts on the pump cover, which also houses a removable check valve.*

Troubleshooting diaphragm carburetors

Carburetors cause a disproportionate amount of trouble, but by no means are the only source of problems. Before rushing to judgment

- Drain the fuel tank and add fresh premix with the correct amount of lubricant.
- Replace the fuel filter.

Throttle body

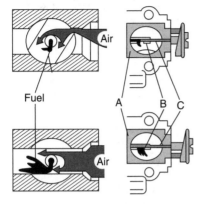

FIG. 4-25. *Barrel-valve carburetor operation.*

- Clean or replace the tank vent valve.
- Visually inspect fuel lines for leaks and hardening.
- Verify that the engine generates at least 90 psi of compression during open-throttle cranking.
- Check ignition output.
- Replace the spark plug with a correctly gapped and known-good plug of the same type as originally installed.
- Service the air filter. Clean foam filters, and prefilters; replace paper main filters.

Table 4-2 is a general guide, listing the effects of carburetor malfunctions upon engine behavior. Note the degree of overlap—any of a dozen carburetor faults have the same or similar effects upon engine performance. Within reason, try to cover all bases by changing out each of the parts supplied in the repair kit.

Pressure tests   Readers who intend to do much carburetor work should purchase a Walbro PN 500-500 tool kit, available from factory distributors. The kit includes cape chisels and punches for dealing with Welch plugs, a slide hammer, and, most importantly, a PN 57-11 pressure tester. Any of these tools can be purchased separately. Other sources of pressure testers include snowmobile and motorcycle dealers. If you have an automotive cooling system leak tester, you can adapt the hose connection to mate with the carburetor inlet fitting.

Metering-system function   The first priority is to determine if the metering system, the source of most problems, functions. Connect the tester to the carburetor inlet fitting and pressurize the circuit to 6 or 7 psi, the equivalent of fuel-pump pressure. The gauge should hold steady (Fig 4-26).

Table 4-2
Diaphragm carburetor malfunctions and their effects upon performance

Carburetor faults	Effects on performance
Improperly adjusted low-speed mixture needle	Hard starting Refusal to idle Stumble upon acceleration Engine quits when closing throttle
Improperly adjusted high-speed mixture needle	Will not run at WOT* Loss of power Quits when closing throttle
Plugged fuel-tank vent	Hard starting Shutdown after a few minutes of operation Refusal to idle Hesitation or refusal to accelerate Engine will not run at WOT Loss of power
Defective fuel line—clogged or leaking air and/or fuel Clogged fuel passage in carburetor body Clogged fuel screen	No or hard starting Refusal to idle Hesitation or refusal to accelerate Engine will not run at WOT Loss of power
Pump failure—loose cover, leaking gasket, defective diaphragm, blocked fuel channels, clogged fuel screen (when fitted), clogged or leaking pulse line	No or hard starting Refusal to accelerate Quits when closing throttle Will not run at WOT Loss of power
Metering system failure—hardened or leaking diaphragm, worn or sticking inlet needle, leaking diaphragm gasket, loose metering chamber cover	No or hard starting Hesitation or refusal to accelerate Quits when closing throttle Will not run at WOT Loss of power
Metering lever set too high and/or distorted lever spring	Overly rich mixture Idles with mixture adjustment screw closed Fuel drips from air horn Quits when closing throttle Erratic idle
Metering lever set too low and/or distorted lever spring	Excessively lean mixture Will not run at WOT Loss of power Hesitation or refusal to accelerate

Table 4-2
Diaphragm carburetor malfunctions and their effects upon performance (*Continued*)

Carburetor faults	Effects on performance
Defective discharge-nozzle check valve (butterfly-throttle carburetors)	Erratic idle Hesitation or refusal to accelerate Quits when closing throttle
Air leak downstream of venturi—carburetor-mounting flange, severe wear on throttle shaft, crankshaft seals	No or hard starting Excessively lean mixture Erratic or no idle Hesitation or refusal to accelerate Will not run at WOT Loss of power Performance improved with choke closed

*WOT = wide-open throttle

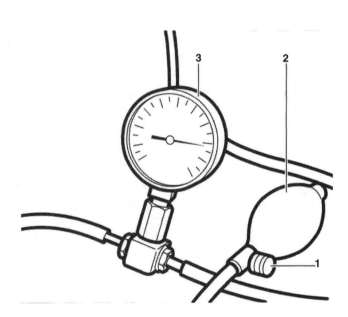

FIG. 4-26. *Pressurize the carburetor through the fuel inlet fitting to 6 or 7 psi. Failure to hold pressure indicates an inlet needle that does not properly seat or a leaking fuel pump diaphragm. A STIHL PN 0000 850 1300 can also be used to test the integrity of the crankcase.*

Close the choke or insert a rag into the carburetor air horn to gain vacuum assist. Pull the engine through several times and note how the gauge responds. If all is well, the gauge needle will flicker as the inlet needle opens and reseats. If the gauge holds steady, the metering system has failed.. Expect to find a stuck inlet needle, a defective diaphragm, or a leaking metering-chamber gasket.

Pop-off and reseat pressure measurement It can be helpful to quantify inlet-needle response. With the tester connected to the carburetor inlet fitting, slowly increase pressure. At some point not always well defined by the literature, pressure will abruptly drop as the inlet needle unseats, or pops off. The reading should then stabilize as the needle returns to its seat.

Ultralight pilots, whose lives depend on their engines, spend hours tinkering with lever adjustments and springs to obtain consistent pop-off and reseat pressures. For less critical applications pop-off pressures can vary by more than 10 psi for otherwise identical carburetors. Reseat pressures are also a little vague, but we need something on the order of 10 psi to close the needle against pump pressure.

Immersion tests If the carburetor fails to hold pressure, immerse it in clean solvent. Bring pressure back up to 7 psi and trace any bubbles back to their sources. If, for example, bubbles stream out the vent hole in the metering-chamber cover, the diaphragm leaks. Remove the cover and repeat the test. There should be no bubbles from the inlet needle until it pops off. Some carburetors may exhibit a few random bubbles from the pump housing without ill effect.

No start Refusal to start can nearly always be laid to insufficient fuel delivery. The spark plug tip will remain dry after extended cranking. A pressure tester will show if the inlet needle unseats, but its good practice to spray a small amount of carburetor cleaner into the spark-plug port. If the engine runs and quits after a few seconds, you can be confident that it suffers from fuel starvation. The fact that the engine runs at all certifies the ignition system.

But there could be a problem with fuel ingestion. In order to verify that the engine can, as it were, feed itself, spray carburetor cleaner into the air horn.

Failure to ingest carb cleaner means that we have a major air leak somewhere downstream of the carburetor. Check carburetor hold-down screws, the mounting flange gasket, and, if it comes to it, the crankcase seals as described in Chap. 6.

Warning: Confine the spray to the carburetor throat and replace the filter element, which acts as a spark arrestor, before cranking. Carburetor cleaner is extremely flammable.

No gasoline delivery can have several causes. If you have not already replaced the fuel filter, replace it now. The next most likely culprits are leaking fuel and/or fuel-pump impulse lines. With plastic lines one must intuit the presence of an air leaks by how hard the lines are and by their age. When in doubt, replace them.

Check the impulse circuit for clogs, which are most likely to develop at the engine casting. Carbon collects in the crankcase side of the port. Remove the carburetor, squirt a little oil into the impulse port, and crank. If the port is open, oil will spill out.

If you do not have a carburetor tester pump, replace the inlet needle and seat as a matter of course. Do the same for the metering and pump diaphragms.

Refusal to idle   Refusal to idle can mean that the engine dies when the throttle is closed or that a closed throttle does not bring rpm down. If the latter is the case, look for

- Maladjusted idle stop screw, also known as the idle rpm screw. Refer to the "External Adjustments" section below.
- Maladjusted throttle cable. The cable should be slightly slack with the throttle resting against the idle stop screw.
- An obstruction in the low-speed circuit, usually at the jet or at the discharge ports adjacent to the throttle blade. Backing out the low-speed adjustment needle compensates for a partial blockage. Removing the needle and squirting carburetor cleaner into its boss can sometimes cure more serious blockages. .

Caution: Blowing out assembled carburetors with compressed air is a prescription for trouble. Pipe cleaners can be used to clear main fuel and air passages, but jets and discharge orifices must be cleaned chemically.

A leaking main nozzle check valve inhibits idle by bleeding air into the low-speed circuit. In theory, one can test this and other check valves by blowing on them through a tube cut square to make good contact with the carburetor body. A functional check valve opens to pass air in one direction and closes in the other direction. It's good practice to replace check valves, assuming that the parts are available and that disassembly can be accomplished with discretion.

Erratic idle   The most common cause of erratic idle is a low-speed mixture needle adjusted to the point of either rich or lean drop-off. Other possible causes are a metering-diaphragm lever set too high, a distorted or incorrectly sized metering spring, or a leaking nozzle check valve.

Many idle-related problems come about because fuel vapor drops out of suspension and coalesces on the intake pipe and crankcase walls. Some puddling cannot be avoided, since the swirl imparted to the charge in the crankcase coats the crankcase walls with fuel. The motion of the crankshaft and connecting rod has a similar effect. About all that can be done to alleviate the situation is to adjust the low-speed mixture screw as lean as decent acceleration permits.

Some two-strokes employ air-vane governors that, when fitted with the wrong or distorted springs, cause the engines to hunt, that is, to gain and lose rpm in a rhythmic fashion.

Poor acceleration   A stumble or flat spot during acceleration indicates that the mixture is too lean. If a conventional (non-wick) accelerator pump is fitted, blipping the throttle should send a stream of fuel into the carburetor bore. Carburetors without accelerator pumps need a rich low-speed mixture, even at the expense of a smooth idle. Back out the low-speed mixture screw as required. A stiff metering spring or a metering lever adjusted too low also results in lean mixtures and hesitation upon acceleration.

Poor high-speed performance   A ragged exhaust note, loss of power, or failure to attain governed rpm generally come about because of insufficient fuel delivery. Look for gum deposits on the main jet, a sticky main nozzle check valve, soft fuel lines that collapse under the increased draw at high speed, and leaking or hardened diaphragms. A faulty tank vent can produce the same symptoms as can restricted exhaust ports.

Check the integrity of the carburetor flange gasket and, if all else fails, the crankcase seals.

Servicing diaphragm carburetors

Once you made a thorough diagnosis, repairs are anticlimatic. But do not expect miracles. Nothing can be done for carburetors like the one in Fig. 4-27.

FIG. 4-27. *Water damage has consigned this carburetor to the scrap barrel.*
Robert Shelby

The next priority is to obtain a repair kit. Some engine makers refuse to catalog these kits, preferring to sell new carburetors instead. But carburetors are off-the-shelf items, used on a variety of engines. Shop around and you can find the parts.

Gasket kits contain diaphragms and the necessary gaskets. Repair or overhaul kits include, or should include, an inlet needle, lever spring, lever, check valves, low-speed and (when fitted) high-speed adjustment needles, Welch plugs, and an inlet screen. These are all the parts normally needed. Figure 4-28 describes areas of concern.

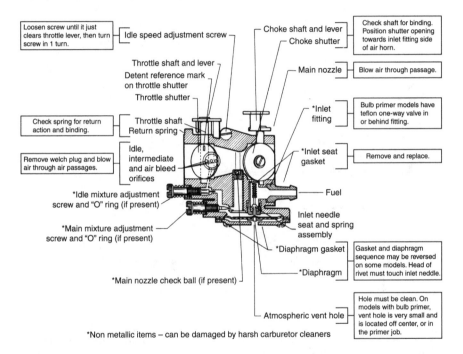

FIG. 4-28. *Repair guide for small Tecumseh diaphragm carburetors has general application.*

Keep a notebook handy to record the routing of fuel lines, the depth of any Welch or cup plugs that you remove, and the gasket/diaphragm sequence.

Cover the work table with paper and lay out the parts in sequence of disassembly. Clean passageways and orifices with an aerosol carburetor cleaner, Q-tips, and pipe cleaners. As the writer can attest, clearing jets with a wire torn from a wire brush upsets the calibration.

Warning: Aerosol spray cleaners contain toluene and other carcinogenic solvents. Work in a well ventilated area, wear gloves, and protect your eyes.

The proliferation of plastic parts and problems associated with safe disposal pretty well eliminate the use of immersion cleaners. Consider products such as Bendix Econ-Clean as a last, desperate resort. Do not allow a carburetor to soak in cleaner for more than 15 minutes and rinse thoroughly with water. A liberal coating of WD-40 (WD stands for "water displacement") prevents corrosion.

Welch plugs   There are times when one must remove Welch and expansion cups for access to internal circuits, but do not rush into the job. Figure 4-29 shows correct and incorrect ways to attack a Welch plug.

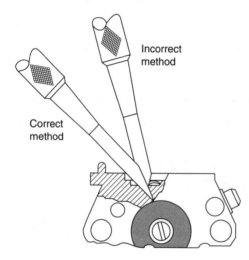

Incorrect
method

Correct
method

FIG. 4-29. *The Walbro PN 500-500 tool kit includes a cape chisel for removal of Welch plugs and correctly sized drivers for plug installation.*

The chisel goes in at a shallow angle to avoid scarring the carburetor body. Note the depth of the plug and puncture it in the center, well clear of its aluminum boss. Small expansion cups can be drilled and extracted with a sheet-metal screw and Vise-Grips.

Coat the edges of replacement Welch and expansion cups with fingernail polish and, using the appropriately sized pilot, tap the plugs into place.

Lever adjustment   As a very general rule of thumb, most metering levers are adjusted to stand flat or just below the surrounding metal. Zama, for example, calls for the metering needle to be 0 – 0.3 mm (0 – 0.012 in.) below the carburetor body. A depth gauge can be used in the absence of the proper factory gauge (Fig. 4-30). You can download a full-sized pattern of the Walbro 500-13-1 gauge at http://www.wind-drifter.com/technical/WalbroGage.pdf.

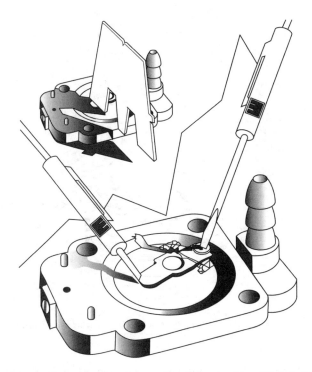

FIG. 4-30. *Walbro throttle levers adjust with a factory gauge. While holding a small screwdriver against the needle—just hard enough to stabilize it—pass the appropriate gauge over the lever. Adjusted correctly, the lever exerts an almost imperceptible drag on the gauge.*

If the lever is too high, bend down on the free end. If too low, pry up on the end , being careful not to damage the needle and seat in the process.

Float-type carburetors

Float carburetors give more precise control over the mixture than diaphragm types and cost no more to manufacture. But float mechanisms are gravity-sensitive and cease to be reliable at angles of 30° off the horizontal. Consequently, their use is limited to motorbikes, snowmobiles, outboard motors, and other applications that, hopefully, stay upright.

Float mechanism   As the level of fuel in the float bowl drops, the hinged float falls away from the inlet needle. Fuel flows past the needle and into the bowl at a much higher rate than the engine can use. The level in the bowl

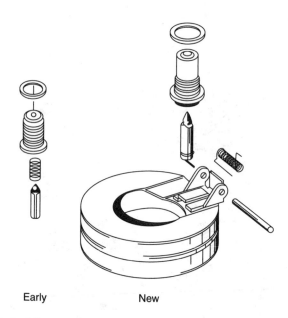

Early   New

FIG. 4-31. *Typical float mechanism, shown with early and late needles.*

rises, lifting the float and forcing the needle against its seat. The primitive, toilet-tank appearance of these valves belies their sophistication. At wide-open throttle the needle seats and unseats 200 times a second.

Figure 4-31 illustrates a "doughnut" float and two versions of the needle and seat. An old-style needle, shown on the left, is made of chrome steel and mates against a brass seat threaded into the carburetor casting. The needle on the left has a Viton tip that, like an automobile tire, flexes to reduce wear. Other carburetors go the opposite route with a Viton seat and steel needle.

In order for the carburetor to function, the float bowl must come under atmospheric pressure. Some bowls vent through an external hose, routed high to allow the engine to tilt without spilling fuel. Others draw air through an internal passage that opens to the air filter.

Float-type carburetors fall into two groups distinguished by the style of throttle.

Butterfly throttle   Most American-made float carburetors have a butterfly throttle, as shown in Fig. 4-32. If we disregard the float mechanism, operation is similar to diaphragm carburetors.

Servicing   Figure 4-33 lists service requirements.

    Needle and seat   Parking a moped or other gravity-fed engine without turning the fuel tap off encourages flooding, since few inlet needles make a

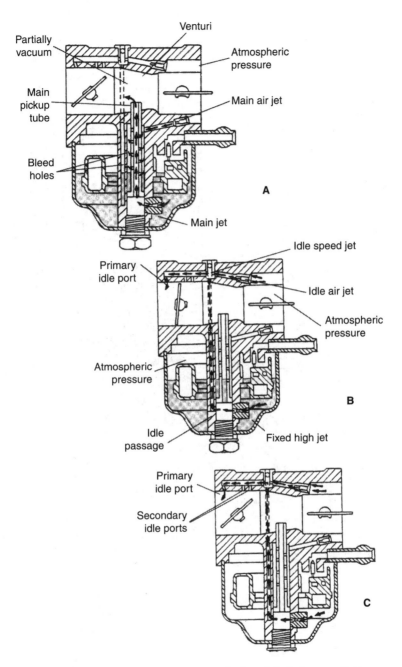

FIG. 4-32. *Float-type carburetor operation. At high speed, fuel enters the bore through the main pickup tube or, as other manufacturers would have it, the main nozzle (A). An air bleed atomizes the fuel prior to discharge. At idle, fuel enters the bore via the primary idle port (B). The (almost always) adjustable idle speed jet and the idle air jet control low-speed mixture strength. As the throttle opens wider, the secondary, or off-idle, ports come into play (C).*

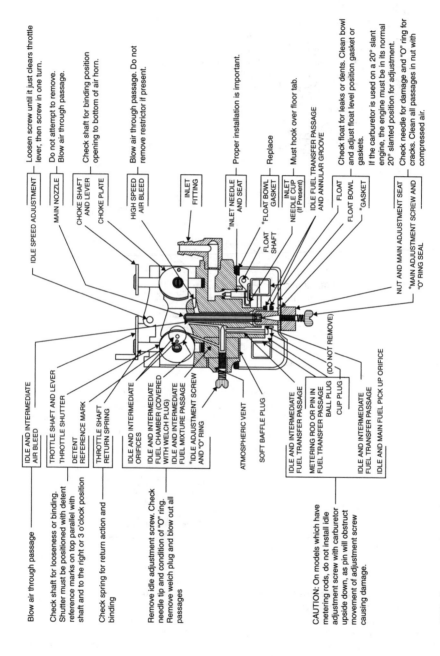

Blow air through passage — IDLE AND INTERMEDIATE AIR BLEED — Loosen screw until it just clears throttle lever, then screw in one turn.

Check shaft for looseness or binding. Shutter must be positioned with detent reference marks on top parallel with shaft and to the right or 3 o'clock position — TROTTLE SHAFT AND LEVER / THROTTLE SHUTTER

— DETENT REFERENCE MARK

Check spring for return action and binding — THROTTLE SHAFT RETURN SPRING

IDLE SPEED ADJUSTMENT

MAIN NOZZLE — Do not attempt to remove. Blow air through passage.

CHOKE SHAFT AND LEVER — Check shaft for binding position opening to bottom of air horn.

CHOKE PLATE

HIGH SPEED AIR BLEED — Blow air through passage. Do not remove restrictor if present.

INLET FITTING

*INLET NEEDLE AND SEAT — Proper installation is important.

FLOAT SHAFT / *FLOAT BOWL GASKET — Replace

INLET NEEDLE CUP (If Present) — Must hook over float tab.

IDLE FUEL TRANSFER PASSAGE AND ANNULAR GROOVE

FLOAT — Check float for leaks or dents. Clean bowl and adjust float level position gasket or gaskets.

FLOAT BOWL

*GASKET — If the carburetor is used on a 20° slant engine, the engine must be in its normal 20° slanted position for adjustment.

NUT AND MAIN ADJUSTMENT SEAT

*MAIN ADJUSTMENT SCREW AND "O" RING SEAL — Check needle for damage and "O" ring for cracks. Clean all passages in nut with compressed air.

IDLE AND INTERMEDIATE ORIFICES

Remove idle adjustment screw. Check needle tip and condition of "O" ring. Remove welch plug and blow out all passages — IDLE AND INTERMEDIATE FUEL CHAMBER (COVERED WITH WELCH PLUG) / IDLE AND INTERMEDIATE FUEL MIXTURE PASSAGE / *IDLE ADJUSTMENT SCREW AND "O" RING

ATMOSPHERIC VENT

SOFT BAFFLE PLUG

IDLE AND INTERMEDIATE FUEL TRANSFER PASSAGE

METERING ROD OR PIN IN FUEL TRANSFER PASSAGE / BALL PLUG / CUP PLUG (DO NOT REMOVE)

IDLE AND INTERMEDIATE FUEL TRANSFER PASSAGE

IDLE AND MAIN FUEL PICK UP ORIFICE

CAUTION: On models which have metering rods, do not install idle adjustment screw with carburetor upside down, as pin will obstruct movement of adjustment screw causing damage.

FIG. 4-33. *Developed for Tecumseh carburetors, this illustration has application for other float-type units.*

reliable, long-term seal. But most flooding occurs spontaneously, either because the needle sticks in its bore or because a speck of dirt prevents the needle from sealing. This type of flooding can be sometimes corrected without removing the carburetor from the engine. If no fuel cutoff valve is present, squeeze the flexible fuel hose shut with Vise-Grips. Remove the float bowl and verify that the needle moves freely. Momentarily open the fuel line. The rush of incoming fuel clears the obstruction.

Warning: Allow plenty of the time for the engine to cool before opening the float chamber, work outdoors, well clear of potential ignition sources.

Some carburetors use an assortment of springs and spring clips to cushion needle action. Make careful note of the orientation of these parts. A mistake during assembly can cause the float to hang and the carburetor to flood.

Normally inlet needles and seats are replaced as a matched assembly, together with a new seat gasket. Grind a screwdriver blade to fit the wide slot in the seat and tighten securely. In an emergency, solid steel needles can be lapped to fit with valve-grinding compound and patience. Viton-tipped needles are non-repairable.

Viton seats, widely used on American-made carburetors, press into the carburetor inlet fitting. Too little installation force allows fuel to leak around the seat, too much force distorts the needle mating surface. Extract the old seat with a hooked wire and install the replacement part with light hammer taps on a 5/32-in. punch. The crushable, ribbed side of the seat should face down, toward the inlet fitting.

To test for leakage, wet the seat with WD-40 and install the needle and float. Turn the assembly upside down and apply air pressure to the inlet fitting. The needle should pop off at around 6 psi and hold 1.5 psi for at least 5 minutes. If not, replace the seat.

Floats The higher the level of fuel in the float chamber, the richer the mixture. Plastic floats have no provision for adjustment, although the mixture can be leaned slightly by using two gaskets on brass seats. Viton seats offer no opportunity for adjustment. Brass floats have a height adjustment in the form of a tang that bears against the inlet needle.

Float height is determined with the carburetor inverted, so that the weight of the float bears against the inlet needle. Where the measurement is taken varies but most manufacturers express float height as the distance between the inboard edge of the float and the roof of the float chamber. Some would have the cover gasket installed, others do not. Use long-nosed pliers and a small screwdriver to bend the tang. Bending the tang by forcing the float against the needle can distort elastomer seats and needle tips.

Some metallic floats have a second tang that limits drop. Excessive drop makes itself known upon refueling an engine that has run dry. The float pivots downward and, if the angle is excessive, loses enough leverage to bind. The resulting flood is of Biblical proportions. At the other extreme, insufficient drop limits the ability of the carburetor to deliver fuel when tilted off the horizontal.

Instructions accompanying the rebuild kit will include float height and, when appropriate, float drop.

The hollow plastic floats used on American carburetors almost never leak. But sheet-brass floats corrode over time. Test by immersing the float in water that verges on boiling. The heat causes air inside the float to expand. The presence of a leak reveals itself by a trail of bubbles.

Sleeping dogs Like good doctors, mechanics do not go deeper into disassembly than necessary. Unless the carburetor is extremely dirty or has definite symptoms of stoppage, leave the Welch plugs that cover internal fuel passages in place. The same goes for throttle and choke plates, which have easy-to-strip screws and must be aligned properly upon assembly. If you must remove these parts, scribe mark the outboard sides of the plates and note the position of the choke cutout. Use red Loctite on the hold-down screws.

Tuning External adjustments—idle rpm, high- and low-speed mixture—are as described for diaphragm carburetors. Err on the side of rich mixtures, since gasoline is cheaper than pistons.

Slide-throttle carburetors

Slide carburetors are used on motorcycles, water craft, and other applications where customers insist upon performance. At WOT, the slide retracts clear of the bore, offering no obstruction to air flow. The slide works in conjunction with a tapered needle to vary both air and fuel passing through the instrument, a configuration that smoothes the transition between part- and full-throttle. These features—a fully retractable throttle and a venturi that sizes itself to engine demand—explain why slide carburetors are also known as smooth-bore or variable-venturi carburetors.

Mikuni, Dell'Orto, Bing, Walbro, or Jikov slide-throttle carburetors differ only in detail. Basic features are the same (Fig. 4-34).

Operation Table 4-3 provides basic orientation for tuning. But as the following text indicates, there is considerable overlap. For example, the idle jet can continue to flow at full throttle and the shape of the throttle cutaway has a real, but diminishing effect on fuel delivery past half throttle.

Some Mikuni carburetors employ a power jet that can be recognized by the external hose that runs from the float bowl to the air horn.

Cruise and high-speed At 1/4 to 3/4 throttle, the clearance between the needle and needle jet controls fuel delivery with some assistance from the air bled and the cutaway milled on the leading edge of the slide. A clip secures the grooved needle to the slide. Raising the needle richens the

FIG. 4-34. *The throttle consists of a tubular slide with a cutaway on the leading edge. The needle moves with the slide to vary fuel delivery in concert with air flow. A "pontoon" float is pretty well standard, in the tradition of early Amal units.*
Robert Shelby

Table 4-3
Major influences on mixture strength by throttle position

Throttle opening	Major influence on mixture
0–1/4	Idle jet
1/8–1/2	Throttle cutaway
1/4–3/4	Needle position
3/4–full power	Main jet diameter
1/2–full power	Power jet

mixture by increasing flow through the needle jet. One can also experiment with needle jet size and, when available, different needle tapers. Mikuni, the premier manufacturer of these carburetors, offers a range of slide cutouts that influence the amount of vacuum acting on the needle jet at partial throttle and air jets of different orifice diameters.

Beyond 3/4 throttle, the needle retracts out of its jet and fuel delivery becomes progressively more dependent upon the size of the main jet, as modulated by the air bleed. When changing main jets, protect the engine by working one size at a time from over-rich towards lean.

Low speed Air flow past the nearly closed slide does not generate sufficient vacuum for the needle jet to flow at between 0 and 1/4 throttle. Fuel enters the carburetor bore through the idle, or pilot, jet and enters the main bore through the idle discharge port, located just aft of the slide valve. While single-port models are encountered, most carburetors have a second, off-idle port that comes into play as the throttle cracks open wider. Fuel undergoes aeration prior to discharge. The pilot air screw works in conjunction with the low-speed jet to meter fuel. The low-speed circuit remains active across the rpm band on Mikuni-pattern carburetors.

Note that air adjustment screws work in reverse of fuel adjustment screws. Tightening an air screw reduces air flow to richen the mixture.

Of-idle hesitation suggests that the pilot jet is plugged or that the air-adjustment screw is too far backed out.

Cold enrichment As used on Mikuni carburetors and their clones, the cold-enrichment circuit works in conjunction with a lever-operated plunger and a closed throttle to deliver a rich mixture for starting (Fig. 4-35). Intake vacuum draws fuel from the bowl, which then passes through the starting jet into a small mixing chamber for aeration. Raising the brass plunger uncovers the discharge port and fuel, impelled by intake-pipe vacuum,

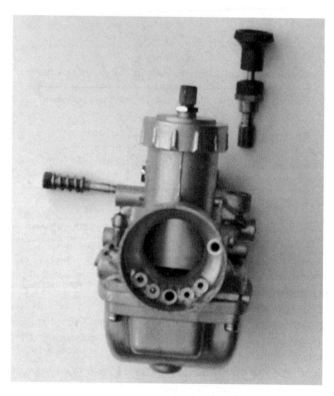

FIG. 4-35. *Raising the brass plunger (shown at the upper right of the photo) activates the starting jet on Mikuni-pattern carburetors.* Robert Shelby

enters the main bore at a point just downstream of the slide. Opening the throttle closes the plunger.

External adjustments—diaphragm and float-type carburetors

Depending upon the application, model, and build date, carburetors have as many as four external adjustments: the throttle-cable anchor, idle rpm screw, and low- and high-speed mixture adjustment needles. The throttle-cable anchor is found on chainsaws and other handheld tools with remote throttle triggers. Adjust to allow some free play in the cable when the trigger is released. At full speed, the throttle should open fully.

The idle rpm screw, also known as the idle stop screw or idle adjustment screw, determines how far the throttle remains cracked at idle. Normally, the factory setting suffices. If the screw has been disturbed, set the idle to factory specifications. Most handheld equipment idle on the upside of 2000 rpm or just under clutch-engagement rpm. Motorbikes and outboard motors can idle a bit slower, but few two-strokes are happy below 1000 rpm.

You might want to invest in a small-engine tachometer, such as the Vibratach, available from Briggs & Stratton dealers for around $40. The German-made instrument senses rpm as the vibration of an extendable wire. Set rpm on the dial, hold the instrument against the engine, and watch as the wire goes into harmonic vibration. There are also a number of electronic tachometers on the market that work from ignition voltage pulses.

Pre-emission carburetors have both low- and high-speed adjustment needles or as they are sometimes called, screws (Fig. 4-36). These needles are spring-loaded to hold adjustment and may make up against O-rings.

The low-speed needle controls the amount of fuel flowing through the idle and transition ports. Turning the needle counterclockwise richens the mixture. The high-speed needle does the same for the main jet. Some manufacturers mark needles "L" and "H," but the location of the needle identifies its function. The needle closest to the engine controls the low-speed mixture. The high-speed needle, whether on the carburetor body, under the float bowl, or behind the air filter, is always upstream of the low-speed needle.

Take a close look at the needles: a visible groove on the tip is normal, but a bent tip or a groove deep enough to hang a thumbnail means that the needle should be replaced. This sort of damage results from forcing needles hard against their seats. Finger-tight is tight enough. A really ham-fisted mechanic can run the needles in far enough to distort the seats. The effect is to make mixtures ultra-sensitive to needle position. Tuning becomes a knife-edged proposition.

In order to meet emissions requirements, manufacturers put restraints on mixture adjustments. Most modern carburetors have fixed high-speed jets

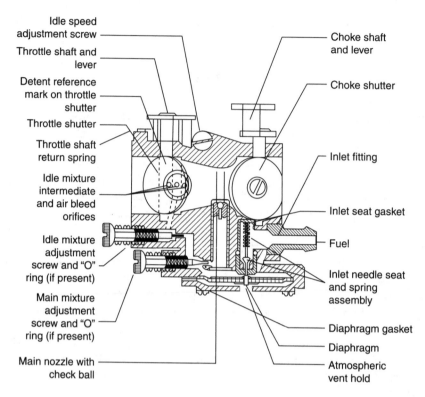

Idle speed
adjustment screw

Throttle shaft and
lever

Detent reference
mark on throttle
shutter

Throttle shutter

Throttle shaft
return spring

Idle mixture
intermediate
and air bleed
orifices

Idle mixture
adjustment
screw and "O"
ring (if present)

Main mixture
adjustment
screw and "O"
ring (if present)

Main nozzle with
check ball

Choke shaft
and lever

Choke shutter

Inlet fitting

Inlet seat gasket

Fuel

Inlet needle seat
and spring
assembly

Diaphragm gasket

Diaphragm

Atmospheric
vent hold

FIG. 4-36. *Tecumseh low- and high-speed adjustment needles do not interchange.*

and fence off low-speed needles behind anti-tampering caps or other kinds of travel limiters. Do not defeat mixture limiters. The EPA has authority to impose a $10,000 fine for modifications that increase emissions of handheld equipment manufactured since 1996.

When emissions are of no concern, carburetor adjustment is a matter of tuning for best power. Adjust the high-speed mixture to the threshold of four-stroking, that is, when the engine skips one or two beats and fires with a bang. Rich mixtures of this kind do no harm to the engine, but can overheat catalytic converters. Lean mixtures are fatal.

The procedure that follows assumes that the carburetor has both low- and high-speed adjustment needles with a full range of travel.

- Start the engine and allow it to reach normal operating temperature.
- Begin with the low-speed adjustment. Turn the needle clockwise in small increments, about an eighth of turn at a time. Allow few seconds

for the effect of each adjustment to be felt. Stop when engine rpm drops off at the lean limit. Note the position of the screw slot.

- Back off the needle in small increments as before. Continue to richen the mixture until engine speed falters at the rich drop-off point. Again, note the position of the screw slot.
- The final adjustment is halfway between lean and rich drop-off points.

Caution: Run lean only as long as necessary to detect the drop-off point. Prolonged lean operation will fry the piston.

- Adjust the idle rpm screw to the specified speed or just below clutch engagement speed.
- With the throttle three-quarters open, repeat the process for the high-speed needle. Set the needle at the midpoint between lean and rich drop-off points. When in doubt, err on the rich side.

Note: Some Tecumseh and Walbro carburetors limit rpm with a governor that complicates the high-speed adjustment. See the "Governor" section above for further information.

- Snap the throttle open from idle. The engine should accelerate almost instantaneously. Stumble, hesitation, or flat spots call for a richer mixture. Most carburetors respond to a slightly richer low-speed mixture, others accelerate more smoothly if the high-speed mixture is richened. Experiment until you find the correct combination.

- Test the engine under load, richening the mixture as necessary.
- To further verify the adjustment, run the engine under normal loads for a half-hour or so, shut off the ignition with the throttle open, and read the spark plug as described in the Chap. 2.

A last word

One can become lost in the complexity of fuel systems, especially when dealing with late-model diaphragm carburetors. Remember that diaphragms cause most problems, followed by inlet needles and seats.

Float-type carburetors give relatively little trouble. Inlet needles and seats are the main culprit, followed by main jets mounted low in the float bowl, where they collect residue and clog.

All carburetors are sensitive to dirty or aged fuel that clogs main jets, inlet screens, and low-speed circuits. Older carburetors can be stripped of their soft parts and immersed in carburetor cleaner for a half hour or so. Many newer carburetors have non-removable plastic parts. For these units, a quick dunk in an immersion-type cleaner is a last option, when all else has failed.

Fuel pumps, whether integral with the carburetor or stand-alone units, require periodic diaphragm replacement. Pumps also malfunction if the crankcase pulse port clogs or if work-hardened plastic lines leak air.

And finally, reading about carburetors is not the same as fixing them. Words, pictures, and troubleshooting charts fall short of the reality. The only way to come to terms with these recalcitrant devices is through the experience of working with them. And if you do much of this work, you will encounter an outlaw, a carburetor that refuses to conform to what we think are the rules. I remember a Harley that ran lean at high speeds, even after drilling main jet oversized.

5

Starters and related components

Rewind starters, also known as recoil or retractable starters, can be intimidating, especially when the main spring gets loose and beats you about the head and shoulders. Nor are factory manuals, most of them anyway, of much help.

After considerable thought, I have arranged this chapter into four sections. The Troubleshooting section gives you an idea of things that usually go wrong. The Overview is just that, a way to become familiar with the lay of the ground. Then there are two types of starters, depending upon the engagement mechanism. Clutched starters are the most complicated and least intuitive. The Eaton and a common Japanese starter of unknown manufacture are in this category, and are discussed in some detail. The information provided should enable readers to repair any of these clutched devices. Ratchet starters which work like ratchet wrenches, are the second basic category. You will find these starters on Craftsmen and other inexpensive equipment, as well as on the sublimely professional Wacker Neuson WM 80.

Troubleshooting

Repairs go more smoothly when you have an idea of what to look for.

- *Sluggish action*   If the starter rope becomes stiff and retracts slowly, assume dirt is the problem. Cold weather that coagulates the grease on the main spring has a similar effect, as does misalignment of the starter with the flywheel cup.

- *Broken cord*   The cord or rope is usually the first part to fail, especially on hard-starting engines that aggravate their operators. Worn rope ferrules contribute to the problem.
- *Failure of the cord to retract*   If the whole length of the cord extends out of the housing, the main spring has either broken or lost its anchor. Partial retraction usually comes about because of a weak spring. Increasing the pretension on the spring sometimes helps, although the surest cure is to replace the spring. Severe starter/flywheel misalignment will also leave the rope dangling.
- *Failure to engage the flywheel*   On clutch-type starters, expect to find a loose clutch retainer screw, a worn or distorted brake spring, or oil on the friction surfaces. Ratchet starters fail to engage because of broken or twisted ratchet springs or rounded-off ratchet tips. These matters are discussed in more detail later.
- *Noise from the starter as the engine runs*   Check starter/flywheel alignment.

Overview

Rewind starters share common features. What is learned by repairing one starter applies, with some reservations, to all. That's the good news. The bad news is that rewind starters contain a spring that, should it escape from its housing, becomes a five-foot-long flail. Wear eye protection and gloves when servicing these units.

Disarming

Before dismantling a starter, the main spring must be allowed to uncoil as much as the housing permits. It will still contain energy. Disarming is accomplished either by detaching the cord handle or by partially disengaging the rope from the sheave.

The handle can be removed by untying the knot or by cutting the cord. Then, using your thumbs as a brake, allow the sheave to unwind. Count the number of turns it makes to be able to apply the same preload upon assembly.

Most sheaves are notched on their outboard sides so that preload can be dissipated with the handle in place. Figure 5-1 illustrates the procedure. Turn the sheave counterclockwise to bring the notch in line with the rope ferrule. Lift the cord out of the notch and allow the sheave to unwind in a controlled fashion. Be sure to count the number of revolutions the sheave makes.

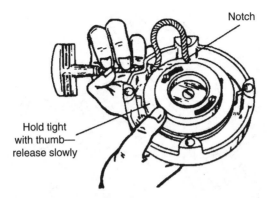

FIG. 5-1. *A notched sheave enables preload to be dissipated with the rope handle attached. Disengaging a few inches of cord from the sheave increases its effective length. The sheave then can unwind with the handle attached.*

Sheave

The sheave, also known as pulley or rotor, is secured by a central screw or by tabs around its rim. Once the hold-downs are removed, the sheave lifts off to expose the main spring. The trick is to lift the sheave without dislodging the spring from its housing. Should this happen, the unconfined spring lashes about uncontrollably, and you will then be faced with the chore of rewinding it.

The inner, or moveable, end of the spring mates with a slot on the underside of the sheave. Slowly turn the sheave to and fro as if tuning a radio. When you no longer feel spring tension, the spring no longer drags against the sheave slot and should remain in its housing as the sheave is lifted off. Carefully raise the sheave: if resistance is felt, stop and start over. I want to warn you again of the importance of wearing safety glasses and gloves when performing operations like this one that can unleash the main spring.

Although one or two manufacturers disagree, sheave bushings need some tiny amount of lubrication.

Main spring

Replace the spring if broken, distorted, or too weak to retract the cord. Otherwise, leave it undisturbed. Replacement springs usually are prewound. Some come housed in a retainer and install as an assembly. Others are secured by a wire clip, reminiscent of the loading clips used on the old M-1 rifle. Align the outboard spring end with the housing anchor and press the spring out of the clip, which is then discarded. STIHL provides dealer

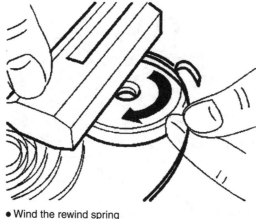

● Wind the rewind spring

FIG. 5-2. *STIHL main spring installation tool (PN 1116 893 4800). The spring winds clockwise into the drum, from the outside inward.*

mechanics with a tool for rewinding the spring (Figs. 5-2 and 5-3). A similar tool—the "EZ Coil Spring Winder"—can be purchased from Week Distributors (telephone: 800/380 3752).

Whether you use a winding tool or install the spring by hand, it is critically important to orient the spring relative to the direction of engine rotation. The spring must wind tighter as the cord is extended. Since the

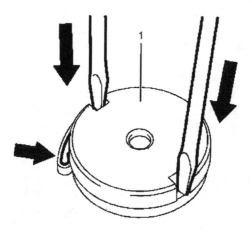

FIG. 5-3. *The spring installs with the assembly block (1) positioned so that the spring loop is captured by the recess in the housing. Once the spring is home, the assembly block is removed.* ANDREAS STIHL AG & Co. KG

normal direction of rotation of nearly all small engines is clockwise, the lay of the rewind spring is counterclockwise. That is, if you trace the curvature of the spring from its outer coil inward, your finger will move in a counterclockwise spiral.

For cold weather operation lightly grease the spring with Molykote PG 54; springs in dusty environments do better with light oil such as WD-40.

Starter cord

The replacement starter cord, or rope, should have the same diameter, length, and weave as the original and it should be made of unalloyed nylon. If the length is unknown, secure the cord to the sheave and wind until the spring coil binds. Allow the sheave to unwind for one or two turns and cut the rope, leaving enough for handle attachment.

Purchasing cord by the engine maker's part number should obviate problems with material. But shops that buy in bulk should be aware that "nylon" cord may contain polyester fiber. This material has only a third of the strength of nylon and wears quickly. To test for polyester, place a sample of the cord in boiling water and add Rit Dye. Nylon will not absorb color, polyester will.

In order to prevent fraying and assist in threading the cord through the ferrule, heat the cut end until it flames. Wipe down the hot area with a rag, tapering the end into a point.

Warning: Allow plenty of time for the cord to cool before touching it. Nylon has almost zero thermal conductivity.

Either of two types of knot can be used (Fig. 5-4) to secure the cord to the handle and sheave. Trim the cord close to the knot and cauterize.

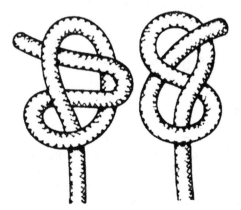

FIG. 5-4. *Nonslip knots are insurance against callbacks.* ANDREAS STIHL AG & Co. KG

Preload

Preload, or the spring tension necessary to fully retract the cord, can be achieved in either of two ways.

Without the cord handle attached:

1. Secure one end of the cord to the sheave. Some manufacturers provide an anchor, others rely upon a nonslip knot.
2. Wind the cord completely over the sheave so that the sheave will turn the flywheel in the normal direction of engine rotation.
3. Install the sheave, anchoring it to the spring.
4. Wind the sheave against the direction of engine rotation the specified number of turns. If the specification is unknown, wind until the spring binds solid and allow the sheave to unwind for one or two revolutions.
5. Hold the sheave against spring tension with C-clamps or Vise-Grips. Now thread the cord through the guide ferrule (also called the rope bushing or eyelet) in the starter housing.
6. Attach the handle.
7. Gently pull the starter through to make certain the cord extends to its full length before the onset of coil bind. Verify that the cord retracts smartly.

With handle attached:

1. Assemble the sheave over the spring.
2. Rotate the sheave, winding the mainspring to coil bind.
3. Release spring tension by one or two sheave revolutions.
4. Block the sheave with C-clamps or Vise-Grips to hold spring tension.
5. With the handle attached, thread the cord through the ferrule in the housing and anchor its free end to the sheave.
6. Release the sheave block and, using your thumbs for a brake, allow the sheave to unwind and wrap the cord around itself.
7. Test starter operation.

Note that the first method must be used when the cord anchors to the underside of the sheave.

Clutch-type starters

These starters evolved from the Jacobsen starter of 1928 that drove through a friction clutch. The clutch housing contains ramps that cam one or more pawls (also called dogs) into engagement with the flywheel. While the housing turns with the sheave, the coupling is less than positive. A friction-inducing spring slows the rotation of the clutch housing relative to the sheave. One end of the spring bears against the housing and the other end against the central hold-down screw or other stationary element. Slowing the rotation of the clutch housing has the same effect as reversing its rotation. The pawls cam into engagement with the flywheel cup.

These starters can be recognized by the coil or hairpin friction spring that bears against the clutch body. American starters may employ a star washer.

Eaton

Eaton starters are standard on American utility and trolling motors. Light-duty models have a single engagement pawl (Fig. 5-5); heavier-duty models use two and sometimes three pawls (Fig. 5-6). Most have a centering pin that rides in a nylon bushing.

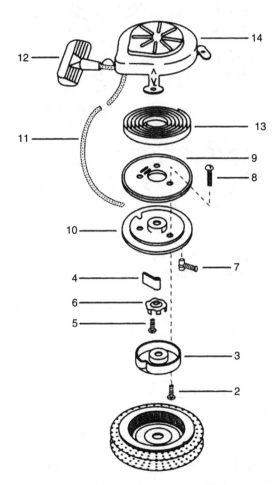

FIG. 5-5. *Eaton light-duty starter of the type frequently specified for Tecumseh two-strokes. This starter is distinguished by a star-shaped brake washer (6) and a two-piece sheave (9 and 10). The main spring (13) installs with the aid of a disposable retainer clip.*

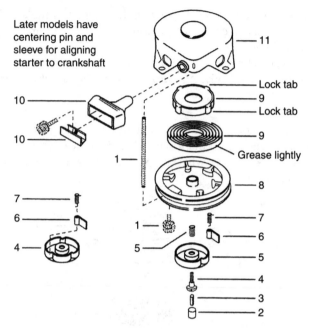

FIG. 5-6. *A slightly more sophisticated Eaton with a single-piece sheave, an optional two-dog clutch (shown on the left), and a cartridge-type main spring and retainer (9). Lock tabs on the retainer OD mean that the spring and retainer remain together as an assembly. Note the absence of a star washer.*

Disassembly

1. Remove the screws securing the starter assembly to the blower housing.
2. Release spring preload. Most sheaves are grooved to allow this operation to be done without detaching the handle.
3. Remove the retainer screw and any washers that might be present.
4. Carefully lift off the sheave while keeping the main spring confined within its housing.

   Warning: Wear safety glasses and gloves during this and subsequent operations.

5. Remove the cord. A hammer-impact tool may be necessary to undo the fasteners on split sheaves.
6. Springs without retainer cups must be unwound a coil at a time from the center out.
7. Clean parts with nonflammable solvent.

Assembly

1. Apply a light film of grease to the sheave pivot shaft and main spring. Do not over lubricate—the brake spring and clutch assembly must remain dry to develop engagement friction. .
2. Install the rewind spring. As shown at 9 in Fig. 5-6, replacement springs for many starters take the form of a preassembled cartridge, which is merely dropped into place. Springs for other models are packaged in a disposable retainer clip. To install, position the spring over the housing anchor pin. The lay of the spring must accord with the direction of engine rotation as described under "Main spring." Cut the tape that secures the spring to the retainer, retrieve the tape segments, and press the spring home into the starter housing. Discard the retainer.
3. Install the cord. How this is done depends upon sheave construction.

   Split sheave (Fig. 5-5):
    (a) Double-knot the cord, cauterize, and install between the sheave halves in the cavity provided.
    (b) Run down the sheave screws, tightening them securely.
    (c) Mount the sheave on its shaft with the main spring engaged. Some Eaton sheaves have an access port so that a small punch can be used to guide the spring end into engagement.
    (d) Using the slot on the sheave OD for purchase, wind the spring with a screwdriver until it coil binds.
    (e) Braking the sheave with your thumbs, allow it to unwind two revolutions. Secure the sheave in this position with a C-clamp or Vise-Grips.
    (f) Guide the end of the cord through the rope ferrule and attach the handle.
    (g) Verify that the rope retracts fully.

   One-piece sheave (Fig. 5-6):
    (a) Install the sheave on the shaft, making certain that the end of the main spring come into engagement.
    (b) Wind the sheave to coil bind and back off enough to align the notch on the sheave OD with the rope ferrule on the starter housing (Fig. 5-7).
    (c) Clamp the sheave.
    (d) Cauterize the ends of the cord and route the cord through the ferrule, over the sheave and into the anchor.
    (e) Knot the anchor-end of the cord.
    (f) Release the sheave, braking it with your thumbs. If all is right, the cord will wind itself over the sheave.
    (g) Test for proper pretension. That is, the spring should retract the cord smartly and retain some resiliency at the limit of cord travel.

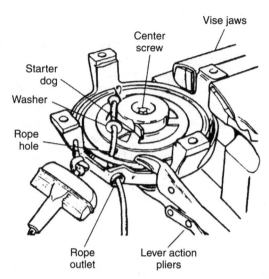

FIG. 5-7. *Installing the cord on a one-piece sheave involves snubbing the sheave with Vise-Grips to hold pretension and inserting the rope from outside of the housing, through the ferrule and into its anchor. Tecumseh starter shown.*

4. Replace any worn or suspect clutch components. Using Fig. 5-5 as the reference, the critical parts are the retainer screw (5), clutch housing (3), star washer (6), and pawl spring (7). As shown in Fig. 5-6, some Eaton starters substitute a coil spring for the star washer and employ two or more pawls.

   Clutch problems outright failure of the pawls to extend or partial failure, which results in slippage. Possible causes are

   - *A loose retaining screw* Apply a few drops of red Loctite to the threads and tighten as hard as the slotted screw head permits.
   - *Oil contamination* Clean parts in solvent and assemble dry.
   - *Loss of star-washer "bite"* Because this condition cannot be detected visually, the washer should be replaced whenever the starter is serviced.
   - *Distorted brake spring.* The spring should stand square.
   - *Wear on the clutch housing* While unusual, replacement is sometimes necessary.
   - *Broken or bent pawl-return springs* Engine kickbacks destroy these springs.

5. Pull out the centering pin (when fitted) so that it protrudes 1/8 in. past the end of the clutch-retainer screw. Some models have a nylon centering-pin bushing.

6. Install the assembled starter on the engine, pulling the cord through several revolutions as the hold-down screws are tightened.

Misalignment is signaled by bind and/or failure of the cord to fully retract. If this is the case, loosen the hold-down screws and reposition the starter assembly on the shroud.

Japanese generic

Tanaka, Shindaiwa, and several other Japanese manufacturers use versions of the clutched starter shown in Fig. 5-8. Repair procedures are outlined in the following discussion; for more detail on cord installation refer to the discussion of one-piece sheave Eaton starters.

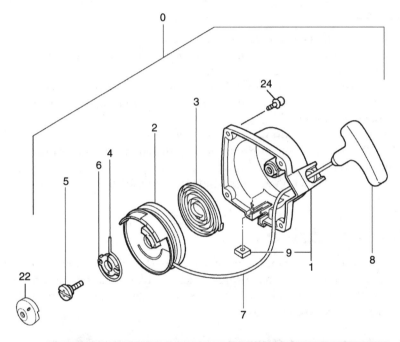

Part number	Description	Qty	Comments
7620304590	Starter Ass'y	1	
7720304580	Body, starter	1	
7740304520	Pulley, starter	1	
7790304520	Spring, recoil	1	
7770304520	Arm, swing	1	
8390304520	Screw, set	1	
7710304520	Stay, pulley	1	
7830304520	Rope, starter	1	
7850304520	Handle, starter	1	
3153334020	Nut, special 5mm	1	
7980304520	Pulley, starter	1	
99464040185	Bolt, hex, 4x18WS	2	

FIG. 5-8. *Tanaka 260 PureFire unit is representative of starters used widely on Japanese engines.*

Disassembly

1. Remove the starter assembly from the shroud.
2. Pull on the handle and, while holding the sheave with your thumb, slip a section of the cord out through the notch provided on the sheave OD.
3. Allow the sheave to unwind in a controlled fashion.
4. With Fig. 5-8 as the reference, remove the central screw (5) and the swing arm (4), and carefully lift the pulley (2) out of engagement with the main spring. Wear eye and hand protection when handling the pulley and spring.

Assembly

1. Clean hard parts in nonflammable solvent.
2. Check for wear or distortion with particular attention to the main spring anchor points and the swing arm.
3. If the main spring has been disturbed, secure its outboard end to its anchor and carefully feed the spring into the housing, winding it counterclockwise.
4. Lightly grease the spring and the sheave post.
5. Install the sheave and cord, preloading the assembly as described previously.
6. Verify starter function and install the unit on the engine.

Ratchet drive

Many handheld tools employ ratchet starters (Fig. 5-9). Instead of extendable pawls these starters have spring-loaded ratchets that bear against teeth on the flywheel hub (Fig. 5-10). Once the engine starts, the ratchets no longer find purchase and coast idly.

Wacker Neuson WM80

Figure 5-11 illustrates the ratchet starter used on the Wacker Neuson ("noy-son") WM80 engine. The main spring mounts "backward," in that its outer end anchors to the sheave and its inner end to the starter housing. And, unlike most, it drives through a single ratchet.

Cord replacement

The combination of a notched sheave and an accessible cord anchor means that the cord can be replaced without removing the sheave (Fig. 5-12). Follow this procedure:

1. Remove the starter assembly from the blower housing.

FIG. 5-9. *Ratchets may pivot off the flywheel as shown here or off the sheave. The hairsprings, barely visible in the photo, are quite fragile.* Robert Shelby

FIG. 5-10. *A toothed wheel transmits starter torque to the crankshaft.* Robert Shelby

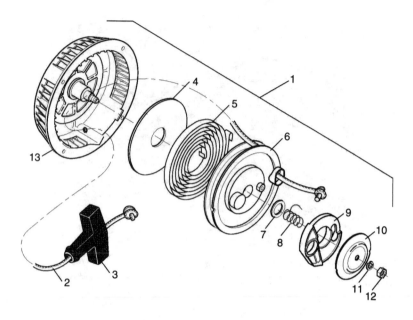

Ref	Description	Ref	Description
1	Starter assembly	8	Spring
2	Rope	9	Ratchet wheel
3	Handle	10	Cover
4	Wear plate	11	Lock washer
5	Return spring	12	Locknut
6	Starter pulley	13	Starter housing
7	Washer	-	--

FIG. 5-11. *Critical areas include the bent main spring ends, the washer (7) which must be installed for the unit to operate properly, and the relation between the ratchet wheel (9) and its spring (8). The main spring should be lightly oiled and the sheave shaft greased.*

2. Lift several inches of cord up through the notch on the sheave. While holding the cord to brake the action, allow the sheave to rotate clockwise until spring tension dissipates.
3. Untie the knot and pull the cord from the sheave.
4. Thread the replacement cord through the sheave and up through the anchor hole. Install the handle and knot both ends of the rope as shown in Fig. 5-12. Trim the end of the cord flush with the knot on the sheave: overhang interferes with starter operation.

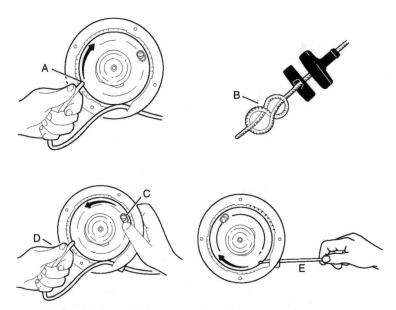

FIG. 5-12. *Wacker Neuson WM cord replacement: disarm the main spring (A), tie nonslip knots on both ends of the cord (B), rewind the cord (C), and verify that the cord extends without coil bind and retracts smartly (D).*

5. Lift several inches of cord through the slot and turn the sheave counter-clockwise two revolutions, guiding the cord as the sheave turns.
6. After two revolutions, disengage the cord from the notch and permit spring tension to wind the cord on the sheave. Repeat this procedure until all 60 in. of cord are on the sheave and the handle makes positive contact against the housing.
7. Install the starter assembly on the blower housing.

Disassembly

1. Remove the starter assembly from the flywheel shroud.
2. Release spring tension as described in the previous section. Using Fig. 5-11 as the reference, undo the locknut (12), lock washer (11), and cover (10). Lift off the ratchet wheel (9), noting its relation-ship to the spring (8). The washer (7) is critical for starter operation.
3. Slowly lift the sheave (6) from the starter housing. If necessary remove the return spring (5), working from the center outward.
   Warning: Wear eye protection and gloves.
4. Remove the wear plate (4).

Inspection

Return spring. The spring ends should be bent 180° as shown in Fig. 5-13(A). Distortion can make it difficult to install the spring in its slotted anchors. Clean the parts in nonflammable solvent.

Sheave. Inspect the sheave for wear at the spring anchor. Clean the mounting stud, grease lightly, and test-fit the sheave. It should turn easily without excessive side play.

Ratchet. If the tip (C) is rounded off, the starter may slip on engagement, and the ratchet should be replaced.

Cord. Replace if worn, oil-soaked, or shorter than its specified 60-in. length. Insufficient cord length can damage the starter.

Assembly

1. Except for the cord, clean all parts with nonflammable solvent.
2. Insert the outboard end of the spring into its anchoring slot (D in Fig. 5-13) on the sheave (F). Wind the spring into the housing being careful to contain the coils within the cavity provided. As mentioned before, protect eyes and hands when dealing with main springs. Once the spring is installed, lubricate with WD-40.
3. Lightly grease the center post and install the wear plate and sheave.
4. Turn the sheave to engage the center end of the spring with the housing.

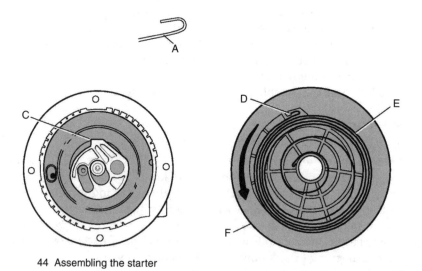

44 Assembling the starter

FIG. 5-13. *Wacker Neuson assembly references.*

5. Install the washer (7 in Fig. 5-11) over the center post and into the recess on the sheave.
6. Mount the ratchet spring and ratchet wheel. These parts go on dry without lubrication. Install the cover, lock washer and locknut. Torque the nut to 6 ft/lb (8 N·m).
7. Install the cord as described previously.
8. Mount the starter on the engine.

Things to remember

Now that we've gone through the major starter types, you should be able to repair any of them. Important points are:

- The main spring winds tighter as the flywheel is turned in the normal direction of engine rotation. If you trace the curvature of the spring from the fixed outer end to inner end where the sheave anchors, your finger would move in a counterclockwise spiral.
- Eaton-style friction clutches assemble without lubrication. Should the clutch slip, verify that friction elements are dry, the center screw is tight, and the drag spring undamaged.
- Failure of the cord to retract can be caused by loss of pretension, a weak main spring or by misalignment between the starter and flywheel hub.

6

Engine service

Major engine repairs are initiated by weak compression, crankcase leaks, and other obvious signs of distress, such as a scored piston or a rattling connecting rod.

Tests

If an engine develops normal compression and the crankcase holds air pressure, the rod and main bearings are probably okay.

Cylinder compression

Handheld engines should develop at least 90 psi of compression for reliable starting. Utility engines are more tolerant: some can be persuaded to start with as little as 60 psi.

To test compression:

- Shut off the ignition or ground the spark-plug lead to protect the coil (and yourself) from high voltage.
- Make up a compression gauge to the spark-plug port.
- With the throttle and choke wide open, crank the engine through three or four times (Fig. 6-1). If compression is lacking, remove the muffler and check the condition of the rings and cylinder as described in Chap. 2, "Troubleshooting."

Crankcase integrity

Worn shaft seals are the major cause of crankcase compression leaks. To determine if a leak is present, we must isolate the crankcase from the atmosphere.

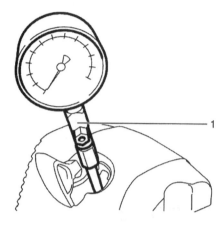

FIG. 6-1. *Small-engine mechanics live by their compression gauges.* ANDREAS STIHL AG & Co. KG

If the cylinder barrel is detachable, replace the barrel with a cover plate. For ease of construction, these plates are usually made of quarter-inch aluminum and sealed with silicone. Drill and tap the plate for a hose fitting and seal off the fuel-pump pulse line, when present. Engines with nondetachable cylinders require two cover plates: one for the intake port and the other for the exhaust. As shown in Fig. 6-2, the intake-port cover incorporates the hose fitting,

FIG. 6-2. *The crankcase and cylinder for this engine share the same casting, which means than both the intake and exhaust ports must be sealed.* Robert Shelby

Vacuum test   A vacuum test replicates conditions during the intake phase, when atmospheric pressure forces seal lips out of contact with the crankshaft. Auto parts stores carry inexpensive hand-operated vacuum pumps, intended for testing emissions hardware. Better quality vacuum pumps can be had from small-engine dealers.

Draw down the crankcase to 7 psi. If pressure remains constant or increases by no more than 4.5 psi over 20 seconds, the seals should be okay.

Pressure test   Careful mechanics double-check the results of the vacuum test by applying positive pressure to the crankcase. This operation requires a 7-psi source of compressed air, which a bicycle pump can provide. Pressure should hold steady for 20 seconds.

Crankcases may also leak across gasketed surfaces or because of casting defects. To check the whole assembly, pressurize the crankcase to no more than 7 psi, immerse it in solvent, and trace any bubbles back to their sources.

Overview

Single main-bearing engines that combine the cylinder and crankcase into a common casting come apart easily (Fig. 6-3). Vertically split crankcases may be a bit difficult to separate, since the main bearings and alignment pins make interference fits with the casting halves (Fig 6-4).

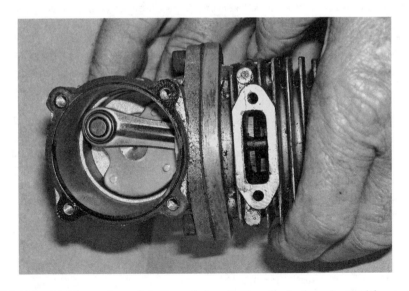

FIG. 6-3. *Cantilevered crankshafts and detachable cylinder barrels simplify repair.*
Robert Shelby

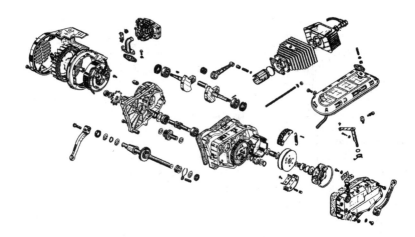

FIG. 6-4. *Vertically split crankcases are standard ware on light motorcycles and moped engines, such as the Morini shown. The resistance of main-bearing fits, sealant, and locating pins must be overcome without distorting the cases.*

Fasteners

Handheld equipment relies upon a mixture of fastener types. Critical components, such as mufflers and cylinder barrels, secure by cap screws, almost always metric. An M8 × 1.0 × 15 screw has a diameter of 8 mm, one thread per millimeter, and a length of 15 mm. The relatively coarse 1-mm pitch is standard for makeup to aluminum.

One also encounters self-threading screws. Shrouding and other plastic parts are secured to plastic with P (Plastoform) screws and to aluminum with DG screws. These specialized fasteners have coarse threads, almost like wood screws, and Phillips, Posi-Drive, or Torx heads.

P and DG screws are not interchangeable and in an ideal world would be installed with a torque wrench. In any event, place the screw in its hole and turn it counterclockwise until it drops down into the existing thread. Failure to feed the screw properly forms a new and weaker thread. It's also important not to interchange binding-head screws (i.e., those with serrations on the underside of the head) with nonbinding screws. Make notes during disassembly and store fasteners and other small parts in labeled sandwich bags.

Table 6-1 lists ball-park torque limits for fasteners that thread into plastic or aluminum. Those designed "M" are conventional metric cap screws and nuts. Note that torque limits are conservative, intended to protect the threads in base material from stripping.

P and DG torque limits vary with the application. For example, 6-mm diameter P screws generally need about 4.5 ft/lb to generate vibration-resistant clamping force. But the same diameter screw, when threaded into a thin

Table 6-1
Approximate torque values for makeup to plastic or light metal

Type and diameter (mm)	Material madeup to	ft/lb	N·m
M4	Aluminum	3.0	4.0
M5	Aluminum	5.9	8.0
M6	Aluminum	8.8	12.0
M8	Steel (crankshaft)	15.0	20.0
P4	Plastic	1.1	1.5
P5	Plastic	3.0	4.0
P6	Plastic	4.4	6.0
DG5	Aluminum	3.0	4.0
DG6	Aluminum	8.8	12.0

blower hose, is tight enough at 3.0 ft/lb. More torque and the threads would strip. It's also true that highly loaded DG fasteners benefit from more torque than indicated in the table.

Less frequently encountered are fasteners that make up to a nut or into a heavy cast-iron or steel component. For these applications, the strength of the fastener determines the torque limit. As shown in Table 6-2, metric bolts come in three grades, the weakest being Grade 8.8. Note that 8.8 bolts shear easily, at least for those of us accustomed to inch-standard fasteners.

Torque tables give one a sense of the lay of the ground. For example, it's useful to know that grade 8.8 metric fasteners are only about half as strong as grade 12.8. It's also useful to know that aluminum threads withstand no more three-quarters of the torque of steel threads. (Actually half or two-thirds of steel torque is safer.). But do not be misled by the emphasis on precise numbers.

Table 6-2
Metric torque limits based on fastener strength

Shank diameter (mm)	Grade 8.8		Grade 10.9		Grade 12.8	
	Ft/lb (in./lb)	N·m	ft/lb (in./lb)	N·m	ft/lb (in./lb)	N·m
M3	0.9 (11)	1.2	1.2 (14)	2.1	1.6 (19)	2.1
M4	2.2 (26)	2.9	3 (36)	4.1	3.6 (43)	7
M5	4.4 (53)	6	6 (72)	8.5	7 (84)	10
M6	7 (84)	10	10	14	13	17
M8	18	25	26	35	30	41
M10	63	86	88	120	107	145
M14	99	135	140	190	169	230
M16	155	210	217	295	262	355

People who work with machinery develop a sense of how tight is tight enough. Torque wrenches come into play only on critical assemblies, such as flywheels and cylinder barrels.

To avoid strip outs when threading into aluminum or magnesium, lubricate the fastener with motor oil, run in the fastener three full turns by hand, and tighten with a 1/4-in. square-drive ratchet wrench. Larger wrenches invite over-torqueing. Draw down fasteners in a criss-cross pattern, working from the center fasteners outward. Torque in at least three increments.

Adhesives and sealants

Critical fasteners, such as those that hold the cylinder to the crankcase, should be secured with a few drops of medium-strength anaerobic adhesive, such as Loctite 242 (blue). Locktite 222MS (purple) is appropriate for 1/4-in. and smaller fasteners. These and equivalent "chemical lock washers" from Omnifit, Prolock, and Hernon develop considerable force as they cure and expand. They should not be used on plastics.

Everyone has a favorite sealant, with RTV (room temperature vulcanizing) silicone a leading contender. However, silicone smears on thick, dissolves in gasoline, and squeezes out with the potential for clogging oil passages. In the writer's opinion, something on the order of Permatex Aviation Form-a-Gasket 80019 is a better choice. This sealant goes on with a brush and withstands temperatures of 400°F (204°C).

Oiling or greasing factory gaskets makes for easy disassembly and, in many cases, allows the gaskets to be reused. But aftermarket gaskets are often made of untreated paper. These gaskets should be doped with Permatex or an equivalent product.

Caution: Remove all gasket and sealant residue prior to assembly. While paint remover can be used to soften old gaskets, the stuff is nasty. Single-edged razor blades of the type sold at hardware stores are a better choice. To avoid gouging the metal, scrape with the blade held perpendicular to the gasket surface.

Crankcase halves often assemble without a gasket, which puts a great deal of responsibility on the sealant. Three Bond Liquid Gasket 1194 is a favorite of the motorcycle fraternity. I've had good luck with Yamabond 5, available from Yamaha dealers.

Philosophy

Basically there are two ways to approach engine work. Pragmatists are content to fix whatever is wrong and move on. The perfectionists among us see engines as collections of raw castings, as diamonds in the rough that need polishing, lapping, and caressing to perform as Nicholas Otto intended. Hopefully, both sets of readers will find this chapter informative.

Housekeeping

Cleaning is a continuous process, carried on as the shrouding, fuel tank, muffler, flywheel, clutch, and other components are removed. Use a nontoxic degreaser and, if possible, perform initial cleaning outdoors. Home mechanics like to use Gunk aerosol, which is a strong alkaline that converts grease and oil to soap in the presence of water. A pressure washer is more appropriate for really dirty tools, such as concrete saws and chipping hammers.

Internal parts should be cleaned with solvent as they come off and cleaned again in fresh solvent before oiling and assembly. The only components assembled dry are electrical gear, tapered shafts (such as used to secure the flywheel), centrifugal clutches, and fasteners treated with Loctite or sealant. Everything else should be lubricated with two-stroke oil prior to assembly.

Covering the workbench with newspaper will reduce dust contamination. You will also need a supply of paper towels and a hand cleaner such as Go-Jo, although laundry detergent works in a pinch.

A bench vise makes a convenient work stand for engines with extended crankshafts. But supporting a functional engine by fixing its crankshaft in a vise may have unpleasant consequences A gentle nudge can bring the engine to life. Then the problem becomes how shut the thing off, spinning several thousand times a minute around its crankshaft. Vulnerable spark plugs, those that stand proud of surrounding metal and plastic, can be disabled with a hammer. Eventually the engine will stop of its own accord as the tank runs dry.

Nor should engines be operated without their shrouds and air cleaners. The same holds for the centrifugal clutches used on portable tools and some motorbikes. For many of these applications, the clutch drum remains with the machine after the engine is removed. Starting the engine without the drum in place sends the clutch shoes flying.

Cylinder head

Detachable cylinder heads secure with four cap screws or long studs that thread into the crankcase. Loosen in an X-pattern. Rusted threads can bind the nuts to their studs with the result that the studs unscrew from the crankcase. If this happens, unthread the nut while holding the stud in a soft-jawed vise. (A pipe wrench or Vise-Grips leave tool marks that act as stress risers.) Once the nut comes off, clean the threads on both ends of the stud and apply a few drops of red Loctite 272 to the crankcase-side threads. Run the stud back into the crankcase using two head nuts, butted hard against each other, to provide torque for assembly.

Examine the underside of the head for ragged carbon streaks that indicate a combustion leak. Gasket failure may be accompanied by a warped head casting or loose hold-down bolts.

Black or dark brown deposits in the combustion chamber mean that temperatures have been normal. Colors fade with increased heat. Two-stroke oils containing calcium oxide stain the deposits yellow or yellow-orange when temperatures have been excessive.

High temperatures result from

- Air leaks downstream of the carburetor. The carburetor-flange gasket is the most likely culprit. Other sources of air include the head gasket, cylinder-barrel gasket, and crankshaft seals.
- Lean carburetor mixtures.
- Clogged cooling fins.
- Detonation. Check for excessive advance on engines with adjustable timing. Detonation can also be caused by using low-octane fuel or excessive amounts of oil in the premix.

Remove loose carbon from the combustion chamber and piston top with a soft wire wheel. A dull knife or hardwood scraper works for stubborn deposits. Try not to mar the aluminum: nicks and gouges are marks of an amateur. Be especially careful of gasket surfaces.

Occasionally one encounters a combustion chamber that looks as if it had been struck with bird shot. This sort of damage results from disintegrated bearings, usually needles from the small end of the connecting rod.

Inspect the cooling fins for damage. Heavy, chalk-like oxidation should be brushed off, and the head protected from further damage with a very light coat of flat black paint. Some perfectionists go to the trouble of filing off casting flash and other irregularities that obstruct air flow. But do not polish the fins, which benefit from the rough, as-cast finish.

Stripped or badly worn spark-plug threads can be renewed with a Heli-Coil insert.

Polishing the roof of the chamber and piston crown yields a small increase in power and makes cleaning easier in the future. The mirror-like finish acts as a thermal barrier to conserve combustion heat. Begin with fine abrasive paper and finish with progressively finer polishing rouge, available from large hardware stores. Of course, all traces of the abrasive must be removed before assembly.

Cylinder heads distort in service (Fig. 6-5), a condition that can be corrected with an "Armstrong milling machine". Tape a piece of 360-grit wet-or-dry abrasive paper to a piece of plate glass or a machine work table. With your hands centered on the casting, move it over the abrasive in Figure-8 pattern.

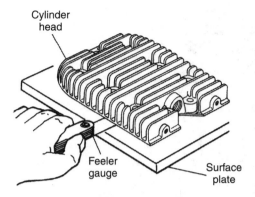

Cylinder head

Feeler gauge

Surface plate

FIG. 6-5. *Measuring head distortion, which on two-cycle engines should be zero or no more than 0.001 in. You can use a piece of plate glass or a machine work table in lieu of a surface plate.*

Oiling the paper results in a smoother finish. The job is done when the entire gasket surface takes on a satiny appearance. The same technique works on warped carburetor flanges.

As a rule, head gaskets should be renewed whenever the head is lifted. However, the copper gaskets fitted to some European engines can be salvaged by annealing, if not too deeply scored. Heat the gasket with a propane torch and quench it in water.

Fiber gaskets sandwiched between steel sheets have a flat and raised side. The flat side goes down, against the cylinder. Fiber gaskets with a ring of light-gauge steel around the combustion chamber install with the wider side of the ring down.

Cylinder heads on some Dolmar chainsaws and European motorbikes incorporate a compression release in the form of a tiny poppet valve. Filling the head cavity with kerosene will show if the valve leaks, a condition usually caused by carbon on the valve stem or, when the valve is remotely actuated, by a sticking Bowden cable.

To install the cylinder head, clean bolt threads with a wire brush to remove all traces of carbon. Torque the head in two or three increments, tightening the fasteners in X-pattern.

Rings and piston

Access to the piston varies with engine design:

- Detachable cylinder barrels, like the one shown back in Fig. 6-4, lift off once the stud nuts are removed. If the barrel sticks, glancing upward blows with rubber hammer will break the gasket seal.

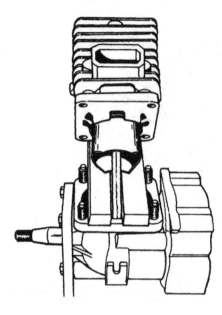

FIG. 6-6. *A slotted wood block prevents the piston from falling against the crankcase as the cylinder is removed and installed. The vertical plate on the left of the drawing is a homemade stand for supporting the engine in a vise.*

Cover the crankcase cavity with a shop rag. It is also good practice to support the connecting rod in a suitably slotted wood block as the barrel is lifted (Fig. 6-6).

• Fixed cylinder barrels with detachable cylinder heads and two-piece connecting rods require that the piston be pushed out from below (Fig. 6-7).

Rings

Note the orientation of the rings and any features that distinguish one ring from another. Two-stroke rings abut against anti-rotation pegs in the piston grooves. Newer engines often have offset pegs that eliminate confusion during installation (Fig. 6-8). Older engines had the top side of each ring marked

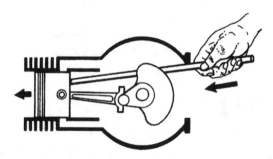

FIG. 6-7. *When the head is detachable and the rod comes apart, drive out the piston with a hammer handle or wooden dowel.*

FIG. 6-8. *If rings were free to rotate, the ring ends would snag on the ports. By looking closely, you can see the locating peg between the ends of the upper ring.* Robert Shelby

for correct installation. The marks vary, but it's safe to say that the side of the ring with any inscription on it goes up.

Ring gap   The clearance between the ring ends is important: if too great, some (the exact amount is in dispute) loss of sealing occurs, and if too small, thermal expansion shatters the ring. It's rare to find problems with ring gaps, but careful mechanics verify the gap.

The "Cylinder bore" section that follows describes how the piston is used to push the ring deep into the cylinder and the gap measured with a feeler gauge. High-rpm portable tools require a minimum gap of 0.006 in. per inch of bore diameter. Less stressed engines are set up a few thousandths of an inch tighter. In the absence of factory data, err on the side of generous gaps.

If the gap is too narrow, support the ring in a vise and lightly caress the ends with a fine-toothed file. Try to keep things square and stone the sharp edges left by the file before using the ring. Sharp edges invite fatigue failure.

Ring replacement   Ring expanders for small-bore engines are difficult to come by; most of us manage with a feeler gauge and long-nosed pliers (Figs. 6-9 and 6-10). Proceed carefully—rings, especially used rings, are razor sharp.

You may want to detach the piston from the connecting rod, as described below under "Pistons," before cleaning the grooves.

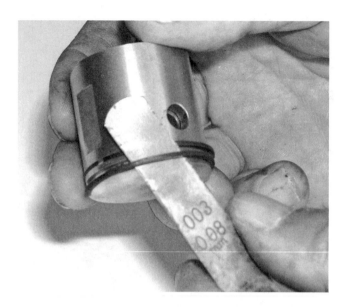

FIG. 6-9. *A feeler gauge can be used like a tire tool to pry rings out of their grooves. When installing the lower ring, the gauge acts as a bridge over the upper groove.* Robert Shelby

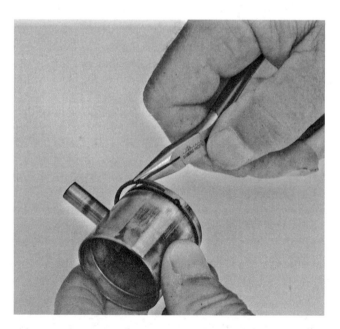

FIG. 6-10. *Long-nosed pliers also help, but be careful not to twist the rings or stretch them open more than necessary.* Robert Shelby

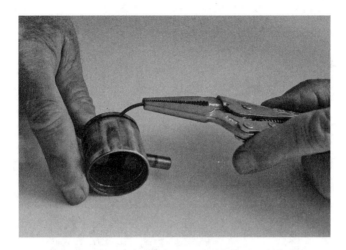

FIG. 6-11. *Cleaning ring grooves prepares the piston to accept new rings.*
Robert Shelby

Break off one of the old rings square and use it as a scraper. To protect your fingers, mount the ring in a file handle or hold it with Vise-Grips (Fig. 6-11). Do not scrape to virgin metal—a thin veneer of carbon on the base of the groove is more tolerable than loss of metal.

The lower, or No. 2, ring slips over the piston crown, past No. 1 groove, and seats with its ends abutting the groove peg. Be careful not to score the piston with the ring ends. Repeat the operation for No. 1 ring.

Piston installation

Most manufacturers chamfer the lower portion of the cylinder barrel to compress the rings as the piston is inserted. Makers of engines with detachable barrels sometimes furnish their mechanics with a ring compressor, although this tool is more of a convenience than a necessity. Oil the piston and rings liberally with two-stroke lubricant and, if the connecting rod is still on the crankshaft, support the piston as shown back in Fig. 6-6. The piston and rod slip off of single-bearing, cantilevered cranks, which simplifies things (Fig. 6-12)

Carefully lower the barrel over the rings. If it binds, do not force the issue. Verify that the rings butt against their locating pegs. A hose clamp, loosely tightened over the rings, can serve as a compressor. The cutout on piston-ported barrels gives some limited access for ring compression (Fig. 6-13).

Once the piston is home, check your work by gently pressing on the rings with a screwdriver from the exhaust port. The rings should flex a few thousandths of an inch and spring back.

FIG. 6-12. *One of the few advantages of having a single main bearing is that the rod slips off the crankpin, making piston installation easy.* Robert Shelby

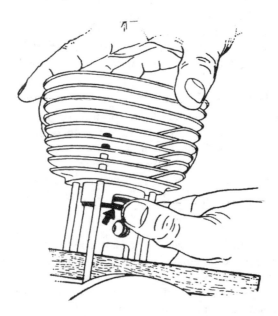

FIG. 6-13. *The cutaway portion on some cylinder liners can be used to compress the rings before they meet the bore.*

Detachable cylinder barrels, especially those held by only two cap screws, tend to come adrift. Apply a few drops of Loctite to the threads and tighten in three increments, drawing down the screws evenly.

Pistons

With good maintenance, a piston should outwear two sets of rings before replacement. Normal wear polishes the piston skirt to a uniform sheen without the deep scores, carbon streaks, or weld splotches caused by overheating or loss of lubrication. The piston in Fig. 6-14 exhibits this sort of damage, as opposed to normal wear. The deep gouge above the ring, probably caused by a flake of carbon, would itself be reason enough to reject the part. Scores on the upper area of the thrust face point to a lubrication failure, either as the result of massive blowby that scoured the oil film or simply because someone neglected to use the right amount of oil in the premix. The way the whole skirt is blackened means that combustion gases leaked past the rings. Yet with all that, combustion temperatures were normal as evidenced by black carbon on the piston crown.

FIG. 6-14. *This sort of damage calls for a new piston and cylinder.* Robert Shelby

Piston removal and installation   Snap rings, called circlips in the trade, anchor the wrist pin to the piston. Should a circlip fail, the pin can drift into the cylinder wall with disastrous consequences. Always use new circlips, positioned in the pin bores their open ends on the long cylinder axis. If the cylinder mounts vertically, with the open ends of the clips should be either up or down. Otherwise, the side forces generated at mid-stroke can loosen the clips.

All modern engines support the wrist pin on needle bearings and many have spacers between the piston and rod end. Small-end bearings may be caged, as shown in Fig. 6-15, or loose. A magnet helps restrain loose needles. When all needles are accounted for, the bearing will be packed without room for another needle.

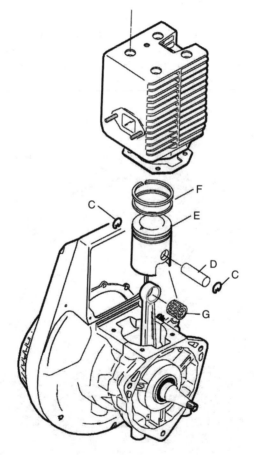

FIG. 6-15. *Wacker Neuson WM80 upper end. The wrist or piston pin (D) rides on a caged needle bearing (G) and is secured by circlips (C). Note that the open ends of the circlips point down.*

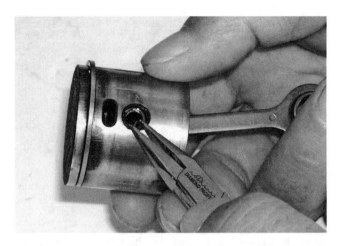

FIG. 6-16. *Circlips with bent tangs are removed and installed with miniature long-nosed pliers. Snap-ring pliers, available from auto supply stores, are used to remove clips with pierced "ears."* Robert Shelby

Conventional circlips, also known as internal retaining rings, are removed and installed with snap-ring or long-nosed pliers (Fig. 6-16). Constant-section, or hook less, circlips as used on the STIHL 4-Mix®, must be pried loose with a scribe. The company makes a clever magnetic tool to aid installation, although a correctly sized punch will work, if you're careful to seat the clip in its groove with the open end up (Fig. 6-17).

Warning: Wear safety glasses when working with circlips.

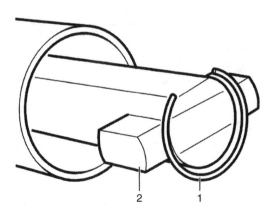

FIG. 6-17. *STIHL's PN 5910 893 1704 employs a magnet (2) to hold the circlip (1) and a sliding sleeve to force it home.*

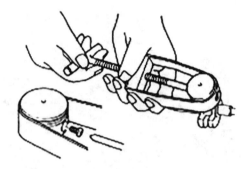

FIG. 6-18. *Tanaka piston-pin extractor. This and other factory tools sometimes appear on Ebay.*

Once the circlips have been removed, the pin can be driven out with a flat punch or, better, pressed out with the appropriate tool (Fig. 6-18). If you use a punch, be extremely careful not to score the piston-pin boss.

After detaching the piston, turn the flywheel to bring the rod to top dead center (TDC) and using Fig. 6-19 as a reference:

- Slide the wrist pin back and forth in its rod bearing. The pin should move easily.
- Attempt to move the pin up and down. If the bearing is good, movement will verge on the imperceptible.
- Try to tilt the wrist pin in its bearing. If there is more than a few thousandths of an inch of play, the needles are tapered and should be replaced.
- Repeat these tests on the big-end bearing. The rod should slide easily along the crankpin, but should exhibit no more than 0.001 or 0.002 in. of up-and-down movement. Movement should be felt as a single phenomenon; if you feel the rod accelerate and halt as you pull on it, bearing clearances are too great.
- And finally, tilt rod on the big-end bearing as was done with the wrist pin and its bearing. If the upper end of the conn rod arcs an eighth inch or more, the rollers have worn conical—that is, they are thicker in the middle than on the ends. Tapered big-end bearings create side forces that can drive the wrist pin into the cylinder wall. And finally, note that the eighth-inch-tilt specification is arbitrary. Less is better (Fig. 6-19).

Rod straightness comes into question when the piston skirt exhibits curved, rather than linear, wear patterns. Figure 6-20 shows the setup, with a pair of machinist's blocks on the crankcase flange and a new wrist pin as reference. Compensate for crankcase distortion by switching the blocks

FIG. 6-19. *An experienced mechanic can evaluate bearing condition by feel.*
Wacker Neuson

from one side of the case to the other. Homemade bending bars are used to
straighten the rod (Fig. 6-21). (Tooth marks left by pliers and pipe wrenches
are fatal to conn rods and other highly stressed parts.)

Thoroughly lubricate the pin and its bearings with two-stroke oil before
assembly.

Unless one end of the wrist pin is capped and the other end open, the
direction of installation is unimportant. However, respect for bedded-in surfaces means that used parts should be assembled as originally found.

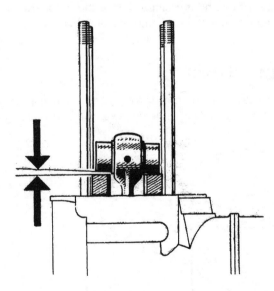

FIG. 6-20. *Check rod trueness with machinist's blocks and a new wrist
pin.* Steyr-Daimler-Puch

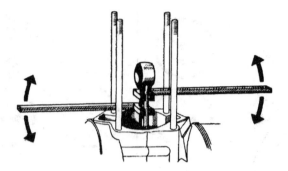

FIG. 6-21. *The lower bar goes just under the bend and the upper bar applies the corrective force.* Steyr-Diamler-Puch

Pistons for high-performance engines generally have an arrow or other reference mark that, upon correct assembly, points to the exhaust port (Fig. 6-22). Pistons without such a mark assemble with the cutaway portion of the skirt adjacent to the intake port. Deflector-type pistons have the steep side of the deflector facing the transfer port. Most loop-scavenged engines have offset rods; reversing the orientation of the piston results in noise and accelerated wear. Doing the same for deflector-type pistons dramatically reduces power.

Shop manuals prescribe that the wrist pin be installed first, followed by the circlips. Most of us find it easier to install the outboard circlip, press the wrist pin into light contact with it, and install the remaining clip.

Cylinder bores

Wipe the bore with a clean shop towel. Severe damage is obvious to the eye, although touch is a far better guide to surface finish. Fingertips can

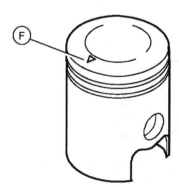

FIG. 6-22. *The arrow on this Wacker Neuson piston crown points to the exhaust port.*

detect irregularities as small as 1 μm (1/25,000 of an inch), or about 100 times smaller than those we can see. Cylinder wear appears as:

- Discoloration that signals loss of ring contact—caused by overheating and distortion.
- Deep grooves along the length of the bore—caused by carbon flakes or broken rings. Replace or, when practical, machine the cylinder oversized.
- Scratches around the exhaust port— if local, these scratches can be tolerated.
- Fine, almost invisible scratches that have the feel of a cat's tongue— caused by abrasive damage. Replace or rebore the cylinder, followed by repairs to the air cleaner or intake gasket.
- Peeled or worn-through chrome or Nikasil coatings on aluminum bores—replace the cylinder.
- Aluminum splatter—chrome-plated and cast-iron bores can be cleaned up with muriatic acid. Flush with water and oil the bore immediately to prevent rust. The underlying cooling and/or lubrication problem must be identified and corrected.

Permissible limits to bore wear and distortion are less rigid than one might suppose. Barring accident, the bore should outlast two pistons or four ring sets. When wear limits are provided, few shops have the precision instrumentation needed to apply them. Most mechanics judge wear by how much the piston wobbles.

Unlike bearing wear that, if allowed to accumulate, results in catastrophic failure, bore wear is relatively benign. Engines with excessive bore wear usually refuse to start before they hurt themselves.

But any deviation from perfectly round and straight cylinders costs power. A few ten-thousandths of an inch can make the difference between winning and losing racing engines. At the other extreme, slow-turning utility engines merely need enough compression to crank. Some of these thumpers run for years with 0.005 in. and more of bore wear.

High-revving portable tools are a special case. Two-inch bores and six-figure rpm put a premium upon precision, especially when Nikasil coatings are involved. For example, the 43-mm bore of Tanaka TBC-600 trimmer has a wear limit of 0.04 mm, or 0.0016 in.

The easiest way to determine bore wear is to measure the cylinder with direct-reading cylinder-bore gauge. To be able to trust the readings, one must invest in a quality Japanese or American tool with a history of recent calibration. Few shops have tools of such precision. Telescoping gauges for use with a micrometer are a less expensive alternative, but require almost superhuman skill to read when dimensions are expressed in tens of thousandths of an inch.

The poor man's approach is to measure wear as ring gap. Using the piston as a pilot, insert the ring deep into the bore in the area that experiences least wear. Measure the gap with a feeler gauge. Repeat the measurement with the ring positioned just below TDC. The difference is, roughly speaking, the amount of cylinder wear.

Old iron

Cast-iron was the wonder metal of the industrial age: inexpensive, easy to pour and machine, and unique in the way it could run against itself without protest. Iron cylinders and cast-iron rings are fully compatible.

Classic engines, such as the Harley-Davidson Knucklehead and Volkswagen air-cooled fours had gray cast-iron cylinders that produce a dull thud when struck with a ball-peen hammer. This is the sound of good vibration dampening.

Unfortunately, the Chinese iron generally available today is not of this quality and lacks homogeneity. It expands unevenly when heated and may crack. If an iron barrel "pings" when struck, it's made of inferior stuff.

But even the best iron is heavy and, relative to aluminum, is a very poor conductor of heat. As a compromise, many utility engines and motorbikes employ aluminum cylinders and thin-walled iron liners, or sleeves. American engines have non-renewable liners, cast integrally with the aluminum "muff." European and some Asian manufacturers machine the cylinder barrel to accept a liner, which is then pressed into place. These liners can be renewed when worn.

The presence of an iron liner can be verified with a magnet, although the ends of the liner are usually quite visible as a ring, about 3-mm thick, around the bore.

Honing Several makers of automotive piston rings have questioned the utility of "glaze breaking," but small-engine mechanics always hone iron liners when new rings are installed. Honing removes small imperfections and roughens the metal to speed the wearing-in process. Chrome rings benefit most.

The tool of choice is a Flex-Hone, a brush with abrasive balls on the end of its bristles and sized 10% smaller than the cylinder bore. Adjustable hones are available for 2-in. and larger bores. Medium-grid (No. 220 or 280) aluminum-oxide stones are appropriate for the chrome-plated rings nearly always used on these engines. Home mechanics may have to use a brake-cylinder hone, which is better than nothing.

Mount the hone in a drill press or low-speed portable drill, lubricate the cylinder with the proper honing oil, and run the hone up and down in the bore. Adjust the reciprocating speed to produce a 45° to 30°

cross-hatch pattern. Keep the hone moving, pausing at the ends of the stroke only long enough to reverse direction. Stop when the cylinder takes on a uniform appearance, a condition that occurs more consistently with a Flex-Hone. Removing the last patches of glaze with an adjustable hone costs more metal than it's worth.

Finish with a mild abrasive such as a Scotch-Brite pad or a piece of No. 600 wet-or-dry paper taped to the hone. A few strokes is all that's necessary to flatten, or "plateau," the peaks left by the hone. This operation also releases some of the entrapped abrasive that would play havoc with rings. But abrasive fragments remain. Scrub the bore with water as hot as you can stand and a strong detergent. Some mechanics favor Tide; others swear by Lava hand soap. Continue to scrub, wiping the bore with paper towels until there is no trace of discoloration. As a final tribute to the persistence of aluminum oxide and to the propensity of iron to rust, scrub the bore with automatic transmission fluid. Chamfer the sharp edges of the cylinder ports with a small stone.

Worn liners can be bored oversize or replaced. But this work requires specialized tools and expertise.

Aluminum bores

Aluminum bores run cooler than iron and, of course, weighs less. And since an aluminum cylinder expands at nearly the same rate as the piston, running clearances can be reduced. But, unlike cast iron, aluminum scuffs when run against itself. If the engine is to live, cylinder bores must be armored with one or another type of coating,

The bores of most quality two-stroke engines are coated with a thin veneer of chromium, applied by an electro-diffusion process that leaves microscopic voids for oil retention. Example include some of the better motorcycles, outboard motors and handheld tools, such as Husky, Elco, Shindaiwa and the Poulan "yellow" series. Low-end manufacturers merely chrome the piston, which then runs against the raw aluminum bore. A chromed piston leaves the cylinder bore at the mercy of the rings.

In 1967 Mahle developed a super-hard coating to combat wear in the Wankel rotary engine. Marketed under the name Nikasil, the coating consists of microscopic silicon-carbide particles imbedded in a nickel matrix. As the nickel wears, carbide becomes the bearing surface. Nikasil is harder than chrome, less liable to peel, and withstands temperatures of 1200°F (648°C). It also dissipates heat better than chrome, and closely mimics piston expansion so that skirt/bore clearances can be almost nil. When the cross-hatching is no longer visible, a Nikasil cylinder has worn past tolerance. Another advantage is that piston rings seal with very little tension, which results in less friction, less heat, and more power.

Nikasil has drawbacks, not the least of which is the cost. Thickness after factory honing can be as little as 0.0015 in., which gives zero protection against the sharp edges of broken rings or ingested solids.

Applications range from Corvette Z-1's and Porche Turbos to quality handheld tools, a category that includes Solo, STIHL, Wacker Neuson, Marurama, Jonsered, and certain Echo products.

Some mechanics attempt to hone Nikasil cylinders when rings are replaced. But, unless they use diamond abrasives, the operation is fruitless. BMW, for one, cautions against honing.

Standard practice is to replace worn cylinders, although U.S. Chrome, Millennium Technologies, and a handful of other companies have the technology necessary to refurbish Nikasil cylinders. But the service is not cheap.

Lower end

The lower end consists of the crankshaft, crankshaft seals, bearings, and, if nondetachable, the connecting rod. Gaining access to these parts is easy when the crankshaft cantilevers off a single main bearing. Nor is the work difficult if the crankcases split horizontally at the main bearings (Fig. 6-23). But vertically split crankcases complicate things.

Vertically split crankcases

Crankcases that split in two vertical slices—each of which holds a main bearing and, for motorcycles, half of the transmission-shaft bearings—pose obstacles to separation:

- A special tool, in the form of a metric-threaded hub puller, is often required to extract motorcycle clutches. Equivalent tools can sometimes be found at bicycle shops.
- Phillips-head screws used by Oriental manufacturers can be difficult to remove even with a hammer-impact driver. If the heads strip, the screws must be drilled out with possible repercussions for the case. Once they're extracted, replace these screws with Torx or Allen screws.
- The adhesive bond formed by the gasket and/or sealant must be overcome, together with bearing and locating-pin fits. There are at least three ways to do this:

1. With luck, a few sharp blows on the crankshaft stub are enough.
2. The least destructive way of overcoming bearing fits, and the way recommended by BMW, is to heat the bearing housing with a propane torch. Keep the flame moving in a circular pattern and stop when the case begins to smoke. Overheating warps the casting and destroys any oil seals that may be present.

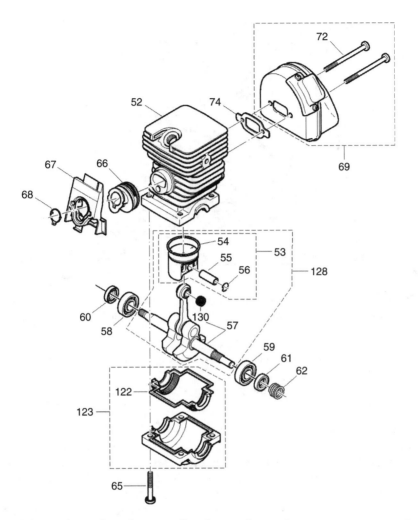

FIG. 6-23. *This Dolmar horizontally split crankcase gives easy access to engine internals.*

3. Another approach is to press out the shaft with a tool that bolts to the crankcase (Fig. 6-24).

4. The locating pins in weather-beaten motors sometimes corrode, resulting in local sticking that cannot be overcome with heat. Typically the cases separate a fraction of an inch and bind. Soak the pins in penetrating oil and wait a few days for the oil to take effect. If that doesn't work, we have to do something radical. While shop manuals warn against prying cases apart, gentle pressure with

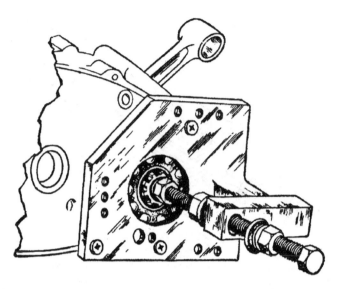

FIG. 6-24. *Joe Bolger, a six-time National New England Motocross Champion, devised this tool for separating Honda cases. Similar tools can be fabricated for other engines.*

large screwdrivers will not necessarily score the gasket surfaces. "Gentle" is the operative word. Work very, very slowly. As the castings separate, keep the gap even. Once apart, replace the pins with new ones.

Inspect the bearing bosses: longitudinal scores reflect the violence of disassembly and can be lived with. Radial scores left by spun bearings require attention. Most mechanics reject such cases, which is tantamount to scrapping the engine. But when the damage is confined to a single spun bearing, the bearing can be secured with Loctite 271 Stud "N" Bearing Fit. Clean the parts to remove all traces of lubricant and apply the adhesive to both parts. The resulting bond requires 16 to 36 N.m torque to shear.

When a gasket was fitted between the casing halves, replace it with a factory part of the correct thickness. Otherwise, bearing "float" will be affected. If no gasket was used, coat the mating faces with Yamabond 5 or an equivalent product.

Crankshaft seals

Replace seals that fail to hold vacuum and/or pressure, and as insurance whenever the crankcase is opened. Stand-alone seals are a mark of quality (Fig. 6-25). Lesser products employ sealed bearings, which compromise reliability and make what would otherwise be a simple repair an exercise in bearing extraction.

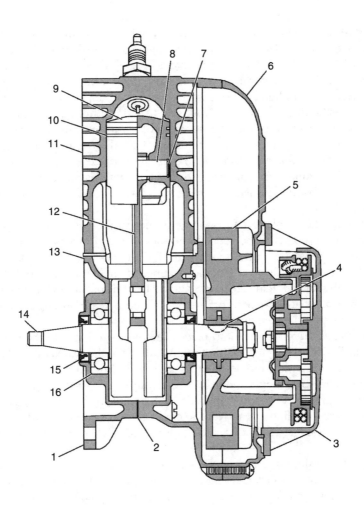

Ref	Description	Ref	Description
1	Crankcase	9	Piston
2	Crankcase gasket	10	Piston rings
3	Recoil starter	11	Cylinder
4	Flywheel key	12	Connecting rod
5	Flywheel fan	13	Cylinder gasket
6	Fan hood	14	Crankshaft
7	Retaining ring	15	Shaft seal
8	Piston wrist pin	16	Crankshaft bearing

FIG. 6-25. *Wacker Neuson cut-off saw engine in cross section. The stand-alone seals (15) can be renewed without disturbing the bearings.*

Stand-alone seals These seals lift out with the crankshaft when the crankcases split horizontally. On other engines, the seals remain in the castings as the crankshaft is withdrawn. Note how deeply the seal is installed. Most are flush or slightly under-flush with the surrounding metal. If the seal cannot be driven out from behind, use a small, blunted chisel, crush the seal inward to break its hold on the casting. Work slowly and carefully, a slip of the chisel can ruin your day. Once the seal is loosened, pry it out with a screwdriver.

The same technique works for engines with their seals accessible from outside of the crankcases with the crankshaft in place. Of course, the presence of the crankshaft complicates things, but with patience one can work around the obstruction. Factory mechanics have access to seal pullers that simplify the job.

Examine the crankshaft for scoring. Some crankcases have deep bosses that permit the seals to be repositioned against an unworn portion of the crankshaft. Polish the contact area with fine wet-or-dry abrasive paper.

Oil seals usually have a flexible coating on their rims to prevent leaks around the seal OD. Plain steel rims should be leak-proofed with a light coat of Permatex Aviation sealant prior to installation. Wipe off any sealant that finds its way to the seal lips.

New seals drive or press into place. The numbered side of the seal goes outboard, toward the installation tool. As a further check, examine the elastomer seal lip, which should have its steep side inward, toward crankcase pressure. On some applications, the seal will have a second lip to prevent ingress from outside of the crankcase. If the crankshaft is still in place, that is, if you are installing a seal over the crankshaft, cover the threads and keyways with a single layer of Scotch tape. Otherwise, the seal lips may be damaged.

The driver should have an OD of slightly less than the seal boss and an ID that concentrates installation force on the seal rim (Fig. 6-26). Automotive tool suppliers can furnish seal drivers in inch and metric sizes, but these tools come in sets and are expensive. A hardwood dowel, with its end cut square, makes a pretty good substitute. Lubricate the seal lips with grease and install the seal to its original depth.

Crankshaft bearings

Most crankshafts run on ball bearings with internal races that protect the shaft from direct contact with the balls. Rod-bearing needles run directly on the crankpin.

Many manufacturers do not offer option of replacing lower-end bearings— when bearings fail, one either scraps the engine or purchases a short block. The discussion that follows is limited to those engines for which replacement parts are available.

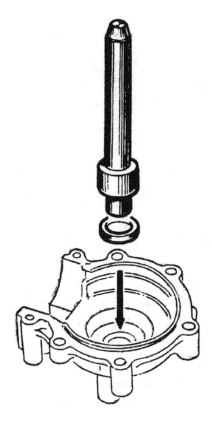

FIG. 6-26. *A hardwood dowel can be substituted for the factory driver shown here.*

Connecting-rod bearings Vintage two-strokes ran their connecting rods on plain bearings. Once in motion, the rods floated on a cushion of oil, which sounds nice. But keeping the rods afloat required oil/fuel mixtures of 16 or even 12:1. Engines smoked like mosquito foggers.

Because needle bearings make rolling contact with their journals, they need very little lubrication. A 50:1 gasoline-oil mixture is plenty and some engines survive on 100:1. Rolling contact also has drawbacks. Any imperfection—a fatigue flake, a skid mark, or a microscopic rust pit— results in rapid failure.

Full-circle rods One-piece rods save weight, at the price of complicating repairs on engines with two main bearings. Access to the crankpin bearing requires disassembly of the crankshaft. While anyone with a 15-ton arbor press can disassemble a built-up crankshaft, putting one back into alignment is an exacting process. Figure 6-27 outlines the procedure, which is easier to write about than do.

Two-piece rods Tecumseh utility engines and some outboards come with two-piece rods, as shown in Fig. 6-28. Split races protect the aluminum rod from contact with the needles that, for reasons that are not clear, may

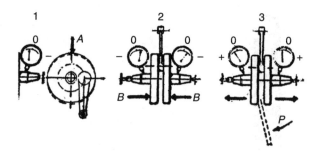

FIG. 6-27. *Shaft misalignments and their cures. If one stub shaft is high, drive its flywheel down with a brass hammer (A). If both shafts are low, squeeze the wheels together (B). And if both shafts are high, pry or wedge the wheels apart.*

be arranged in single or double rows. Match marks on the rod and cap must be aligned for correct assembly and rod bolts torqued to specification.

Main bearings Evaluating the condition of antifriction bearings involves more art than science. The first step is to remove every trace of oil from the bearing. Allow time for the bearing to air dry and turn it by hand. Any roughness or binding means that the bearing verges on failure. If you have access to a new bearing, use it as a touchstone to determine how much radial and axial play is permissible. A sign of doom is an outer race that slips off the balls.

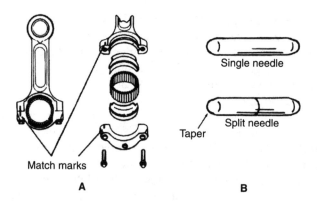

FIG. 6-28. *Two-piece connecting rods with big-end needles are standard on Tecumseh utility and outboard engines. Note that rod and cap match marks must be aligned (A) and that needles should pack solidly around the crankpin. If there's space for another needle, you have lost one. The squared-off ends of split needles face each other (B). Grease or beeswax can be used to hold the needles in place during assembly.*

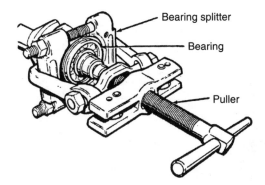

FIG. 6-29. *A bearing splitter extracts some, but not all, main bearings.*

Caution: Do not spin bearings with compressed air. While gyroscopic progression is an interesting phenomenon, damage to the bearing inevitably occurs.

Figure 6-29 shows how a bearing splitter and gear puller collaborate to remove a main bearing. When space restrictions prohibit the use of a splitter, factory tooling (or some facsimile of it) must be used (Fig. 6-30).

The installation setup for Wacker Neuson engines can be used as a model for others (Fig. 6-31). The pusher tube (F) bears against the inner race and the crank support plate isolates the crankpin from bending forces. New bearings go on with their marked sides toward the pusher tube.

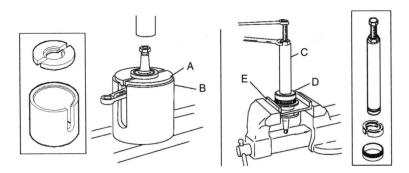

FIG. 6-30. *One of the nice things about quality products is the thought that goes into special tools. To remove a Wacker Neuson main bearing, mount the crankshaft in the PN 0023339 support tube (B) and position the PN 0023338 split ring puller (A) against the bearing. Puller halves may need to be taped into place. The bearing is then pressed out. As shown on the right, Wacker Neuson also supplies tools—the PN 0013288 extractor (C), the PN 0017328 half shell (D), and the PN 0013290 holding ring (E)—that enable bearings to be extracted without an arbor press.*

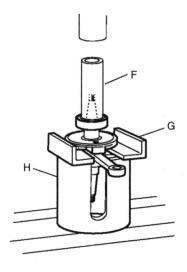

FIG. 6-31. *Bearing installation using Wacker Neuson tooling. Note that the nearside crankshaft web must be supported.*

Another way to install bearings is to create a heat differential between the shaft and the bearing. Since the interference fit on is on the order of 0.001 in., a differential of 150°F (65.6°C) allows "palm-push" assembly. A light bulb placed against the inner race provides sufficient heat, although one must be careful not to exceed 250°F (121°C), the temperature at which that bearing steel becomes reactive. The bearing can also be heated in oil. If you use this approach, support the bearing above the bottom of the container on wire mesh and work outdoors. Carefully monitor oil temperature.

Now that the hard part is done

Once you have the engine back together, check for any "extra" parts on the bench. Verify that the crankshaft turns easily without hard spots or binds. Fill the tank—without adding extra "break-in" oil to the premix—and examine the fuel lines and carburetor for leaks.

Pull the starter cord through a few times. Resistance should increase at mid-stroke and peak as the piston rounds top dead center. Engage the choke and give the cord a vigorous pull. Do not be dismayed if the engine does not immediately start. The problem is almost surely an oil-fouled spark plug. Once the engine comes to life, it will smoke a bit until the oil used during assembly burns off. Run at less than full throttle and under moderate, but varying, loads for the first few tanks of fuel. Stop periodically to allow the engine to cool.

7

Power transmission

This chapter describes how to service clutches, belts, shafts, and other power-transmission components. This information will also be helpful to readers who want to adapt small engines for other uses.

Centrifugal clutches

A centrifugal clutch, the mirror image of a drum brake, uses friction to transmit power. Figure 7-1 illustrates the typical layout, consisting of a pair of shoes restrained by a coil spring. The shoe assembly keys to the crankshaft and is secured by a nut that may have left-hand threads. The drum rides on a needle bearing or, on less expensive machines, a brass bushing (Fig. 7-2). The level of sophistication increases for chainsaws and other heavily used tools (Fig. 7-3). Clutch-in speed varies with the application, but is usually near the torque peak of the engine. Reducing spring tension or adding weight to the shoes lowers the clutch-engagement speed.

When overloaded, a centrifugal clutch behaves almost like a variable-speed transmission: engine rpm drops, the clutch disengages, rpm recovers, and the clutch again engages. On- and off-again engagement does, however, generate large amounts of heat.

Clutch failure is usually the result of overheating from application of greater loads that the engine can manage or that the clutch was designed for (Table 7-1).

Builders of trimmer- and chainsaw-powered bicycles, mini-bikes, and other one-off creations often encounter clutch problems. OEM centrifugal clutches cannot tolerate the loads. Until recently, the only option was an

FIG. 7-1. *Two-shoe clutch typical of garden tools.* Robert Shelby

aftermarket clutch designed for four-stroke engines with crankshaft diameters of between 5/8 and 1 in. One had to bush the clutch down to the size and play with spring tension to match two-stroke torque characteristics. But that has changed: Comet Industries now offers a 3/8-in.-bore centrifugal clutch two-stroke engines of between 0.5 and 2.5 hp.

FIG. 7-2. *When the clutch drum rides on a bushing, frequent lubrication is required.*
Robert Shelby

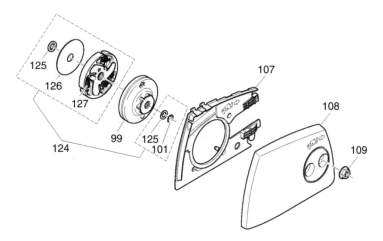

FIG. 7-3. *The clutch for a Dolmar chainsaw is more sophisticated than those pictured previously. Major components are the drum (99) and shoe and spring assembly (127).*

Repairs

Repairs are not difficult, once we realize that most of these clutches secure with left-hand threads. Also, be aware that starting an engine without the clutch drum in place invites serious injury. The shoes, no longer restrained by the drum, become projectiles. And even with the drums in place, clutches have been known to explode. Racing go-karts have their clutches encased in heavy-gauge steel guards.

Table 7-1
Clutch Malfunctions and Their Probable Causes

Symptom	Probable Cause
Failure to disengage	Weak or broken shoe springs
	Idle speed set to high
	Sticking shoe pivots
Slip under load	Oil on friction surfaces
	Worn shoe linings
	Worn drum
	Overloading
Radial (side-to-side) play on drum	Worn drum bearing

Follow this procedure:

1. Remove the engine from the frame. On some designs, the clutch drum remains with frame as the engine is detached.
2. If the hold-down nut must be removed for access, determine the thread direction. Some left-hand-threaded units are marked; others can be identified by close inspection. Remove the nut with a power impact wrench. When hand tools are used, lock the crankshaft with a length of nylon rope fed through the spark-plug port, as illustrated back in Fig. 3-2.
3. Verify that the driven element can be turned by hand. If not, determine the cause.
4. Carefully examine the clutch drum for cracks, heat checks, and scores. Replace as necessary.
5. The shoes may be restrained by an encircling garter spring or, if pivoted, by individual springs (33 in Fig. 7-4). Pry the springs off with a screwdriver (Fig. 7-5). Do not be surprised to find that the springs have lost tension because of overheating.
6. Friction surfaces must be dry, if the clutch is not to slip. A leaking pto crankshaft seal will oil the clutch.
7. Inspect shoes for wear. Heavy-duty shoes may be lined with Kelvar or another non-asbestos friction material.
8. Remove the fastening hardware and lightly grease the shoe pivots.

V-belts

A V-belt transmits power by wedging its angled sides into the pulley grooves. As the belt wears, it sinks deeper into the pulley, or sheave. Eventually the belt rides on the base of the groove, defeating the wedging action. Very little torque is transferred.

Sheaves "dish" in service, a condition that can be detected with a straightedge. Any light between the straightedge and the flanks of the groove is grounds for replacing the sheave. Extreme wear sends the belt into base of the groove.

Adjustment

Verify that pulley shafts are dead parallel and align pulleys with a straightedge so that the belt centers in the grooves.

Belt tension is difficult to generalize about since the correct tension depends on multiple factors, such as the arc of contact on the smaller pulley, belt width, and distance between pulley centers. Too little tension results in belt flap and slippage; too much tension costs power and loads bearings. As a very broad rule, belts for small-engine applications should deflect 1/4 in.

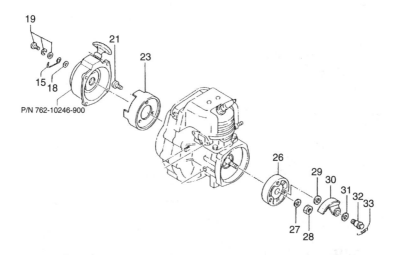

Item	Part number	Description	Qty	Comments
15	461-33111-200	Clamp	1	
18	992-01050-011	Washer, 5	1	
19	994-15050-202	Screw, 5x20/PS	4	
21	994-14060-202	Screw, 6x20S	3	
23	798-10246-200	Pulley, starter	1	
26	347-10243-200	Flange, clutch	1	
27	992-01100-011	Washer, 10	1	
28	991-15100-001	Nut, LH, 10	1	
29	359-10112-203	Washer, clutch B	3	
30	290-10112-802	Arm, clutch	3	
31	358-10112-201	Washer, clutch	3	
32	357-10112-204	Bolt, clutch step	3	
33	342-10205-220	Spring, clutch	3	

FIG. 7-4. *Triple-shoe clutch used on the Tanaka TLE 550 edger. The shoes are lined to improve performance and reduce drum wear.*

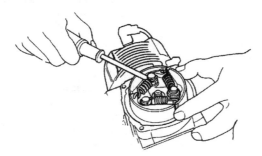

FIG. 7-5. *Springs pry off. Replace the springs if blued, distorted or if the clutch refuses to disengage or engages early.*

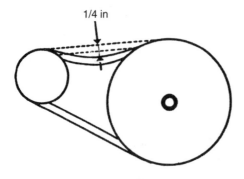

FIG. 7-6. *A quarter inch of deflection should be sufficient for most applications.*

between centers under moderate thumb pressure (Fig. 7-6). Retest after several hours of operation—all V-belts stretch.

Troubleshooting

Table 7-2 lists belt maladies and their causes.

Belt sizing

The 40° included angle on V-belt sheaves is constant, but other aspects of belt design vary. DIN (German Institute for Standardization), ISO (International

Table 7-2
Belt Malfunctions and Their Causes

Belt Symptom	Probable Cause
Excessive wear	Sheave grooves worn
	Sheave misalignment
	Sheave diameter too small for the amount of power transferred
Rough sidewalls	Slippage caused by misalignment or worn sheaves
Break in fabric cover	Prying belt over sheaves during installation
Cracks in base of belt	Insufficient tensioning
	Slippage
Slip burn (localized cauterization)	Slip during starting
Swelling or softening	Oil contamination
Flap	Insufficient tension
	Unbalanced sheaves
Belt turns over in sheaves	Misalignment
	Sheave wear
	Insufficient tension
	Worn-out belt

Table 7-3
V-Belt Nomenclature and Profile

Section	Width (in.)	Height (in.)	Comments
A	0.50	0.31	American "Classic," still used on industrial
B	0.66	0.41	and other heavy-duty applications
C	0.88	0.53	Length expressed as ID B40 = 40-in. ID
D	1.25	0.75	
2L	0.25	0.16	Automotive narrow-profile, less power
3L	0.375	0.22	transfer capability than similar
4L	0.50	0.31	"Classic" belts
5L	0.66	0.375	Length expressed as OD 4L300 = 30-in. OD
3V	0.38	0.32	Narrow profile for use where space is limited
5V	0.62	0.54	Length expressed as OD 5V400 = 40-in. OD
8V	1.00	0.88	
AX	0.50	0.31	Cogged to accommodate small sheaves and
BX	0.66	0.41	for better cooling
CX	0.88	0.53	
3VX	0.38	0.32	Cogged
5VX	0.62	0.54	
8VX	1.0	0.875	
SPZ	0.38	0.32	Metric
SPA	0.50	0.39	Length expressed as OD, ID, or
SPB	0.64	0.51	other variables, depending upon the
SPC	0.87	0.71	manufacturer

Abbreviations: ID = inner diameter, OD = outer diameter.

Standards Organization), and RMA (Rubber Manufacturer's Association) each have their own standards for belt profiles and for the way lengths are expressed (Table 7-3).

Purchase the exact replacement by brand name, especially when dealing with metric belts. Interchange lists are never exact. When dimensions are unknown, measure the sheave groove to determine width and use one of the Internet calculators to arrive at the length.

http://www.csgnetwork.com/pulleybeltcalc.html

http://www.dxpe.com/Gadgets/gadgets/belt_length.asp

Measure the sheave center-to-center distance with the sheaves as closely together as the adjustment permits.

A belt running on fixed pulleys has some ability to multiply torque. The pulleys in Fig. 7-7 have identical diameters. But effective diameters depend upon belt tracking. Under load, the lower side of the belt tenses and the upper side relaxes. The belt buries itself deeper into the driven pulley and moves outward on the drive pulley. Torque is multiplied.

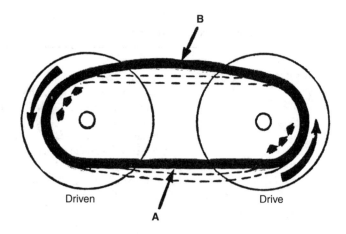

Driven   Drive

A

| **A** Results from resistance of terrain condition, load, etc. | **B** Side is slack and thrown in direction of power line. Result = Improved torque ability |

FIG. 7-7. *Fixed sheaves benefit from self-induced torque multiplication.* Bombardier Ltd.

Belt-driven torque converter

While self-induced torque multiplication is always welcome, the effect is minimal. Serious torque multiplication requires a torque converter. These devices act as centrifugal clutches to disengage the belt at idle and, at higher engine speeds, provide stepless shifting. Ratios blend seamlessly, which gives the feel of steam traction and extends the life of chains, sprockets, and other downstream components.

Belt-driven torque converters are an excellent choice for home builders, who want to give their mini-bikes, scooters, and go karts flexibility without the expense and complication of gearboxes. High-tech versions of these transmissions are used on several automobiles, including Honda and Toyota hybrids.

Operation

Figure 7-8 illustrates driver assembly for the 30 Series Comet. Pulleys consist of two flanges, one of which is moveable. As the drive-pulley flange moves inward in response to engine rpm, the belt climbs in its groove to increase the effective diameter of the pulley. Since the length of the belt does not change, the spring-loaded flange on the driven pulley opens,

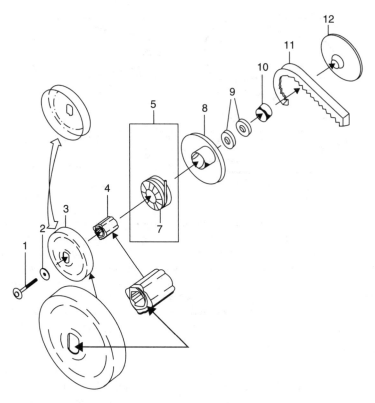

FIG. 7-8. *Series 30 Comet driver assembly for engines developing between 3 and 8 hp. The flyweight and spring assembly (5) cams the moveable flange (8) toward the fixed flange (12) in response to engine rpm.* Go Kart Supply

reducing its diameter. As one pulley becomes larger, the other pulley must become smaller. The inverse relationship between pulley sizes provides torque multiplication.

Comet transmissions also incorporate a torque-sensing feature. As load increases (say, when climbing a hill), the driven pulley cams are closed, which forces the drive pulley to open for an increase in torque multiplication.

At idle, the belt rides loosely on a brass bushing at the base of the drive sheave. In this position it transmits no torque and generates no friction (Fig. 7-9A). As the throttle opens, centrifugal weights cam the pulley's moveable flange inward (Fig. 7-9B). The belt, squeezed between the moveable and fixed flanges, climbs higher in its groove to increase the effective diameter of the pulley. At the same time, the driven pulley opens, making its diameter smaller. Engine torque is multiplied by a factor of three.

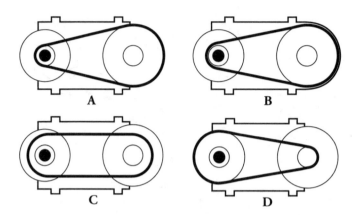

FIG. 7-9. *At idle, the drive pulley opens wide so that the belt rides loosely between the drive flanges, neither transmitting power nor generating friction (A). As rpm rises, the moveable flange moves inward, gripping the belt to initiate power transmission (B). The difference in pulley diameters creates a torque multiplication of about 3:1. Open the throttle more and the belt rises on the drive pulley and sinks deeper into driven element (C). Pulley diameters equalize. At wide-open throttle the drive pulley overdrives its mate, reducing engine revolutions by about 20% (D).* Go Kart Supply

With increased rpm, the belt moves further out on the drive pulley and the driven pulley becomes correspondingly smaller (Fig. 7-9C). Pulley diameters equalize. Should load increase, as during acceleration, the engine slows and the transmission downshifts to a lower ratio.

At high speed, the mismatch between pulley diameters sends the transmission into overdrive (Fig. 7-9D). That is, the driven pulley turns about 20% faster than the drive pulley to reduce engine wear and fuel consumption.

Troubleshooting

Creep at idle can be caused by a belt that is too short or one that is installed wrong. The asymmetric belt used on 30 Series converters should have its flat side next to the engine. Pulleys need to be in dead alignment.

Check that the springs and pins in the flyweight assembly are tight. Any play in these parts prevents the transmission from disengaging. A worn hub or loss of driven-pulley spring tension has the same effect.

Drive belt malfunctions and their cures are described in Table 7-4. To that it should be added that the drive pulley includes plastic buttons that limit how far the moveable flange opens. Wear or loss of these buttons allows the underside of the belt to contact rivet heads on the pulley. The resulting wear is rapid.

Table 7-4
Drive Belt Diagnosis

Symptom	Probable Cause	Comments
Rapid belt wear	Drive belt installed wrong	Series 30 torque converters use
	Wrong belt	an asymmetric belt with the
	Belt too short	flat side toward the engine.
	Pulley misalignment	
	Malfunctioning drive pulley	
	Overloading—climbing hills,	
	dragging brakes, etc	
Belt worn thin in one spot	Locked axle	Clean and lubricate drive
	Drive pulley malfunction	pulley.
	Idle speed too high	
Belt cupped (worn concave on flanks)	Excessive run out on drive pulley	Repair or replace drive pulley.
Belt disintegrated	Excessive engine rpm	
Belt "roll over"	Pulley misalignment	Align pulleys.
	Excessive run out on drive pulley	Repair or replace drive pulley.
	Excessive belt speed	Reduce engine speed.
Cord breakage on belt edge	Pulley misalignment	Align pulleys.
Belt worn unevenly on one side	Pulley misalignment	Align pulleys.
Belt glazed	Pulley malfunction allowing belt to slip	Check drive and driven pulleys.
	Excessive horsepower for converter	Contact Kart Supply for support.
	Oil on belt, often a residue of belt manufacturer's mold release	Clean belt and pulleys.
Wear concentrated on the upper edges of the belt		Replace belt with correct PN.

Erratic engagement is usually caused by sticking flyweights or a binding hub in the drive pulley. Disassemble, clean, and lubricate with a dry molybdenum-based lubricant such as Comet GP-730A. Figure 7-10 indicates lube points. Conventional greases cannot withstand the heat generated by these transmissions.

Adjustments

The tension of the coil spring that holds the moveable flange of the driven pulley against the fixed flange can be varied. The spring anchors in

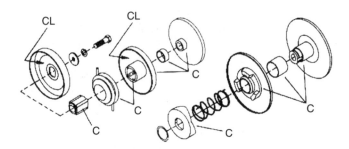

FIG. 7-10. *Series 30 transmission with cleaning (C) and lubrication points (L).*
Go Kart Supply

any of three numbered holes. Hole No. 1 provides least tension and early shifting; No. 2 is the factory default setting; and hole No. 3 is for uneven terrain.

The foregoing discussion owes much to Chet Dowden, president of Go Kart Supply (www.gokartsupply.com) and a long-time Comet dealer. Chet can be reached at 318 925 2224 or by email at gokarts@gokartsupply.com.

Drive chains

The American Standards Institute (ASI) assigns a two-digit code number to roller chains. The first digit describes the pitch, or the spacing between link centers, in eigths of an inch; second digit is "0" for standard chains. Thus, a No. 40 chain has a pitch of 4/8 or 1/2 inch. A No. 41 is a narrow version of the 40. Metric chains are also sized in inches, but the count is based on sixteenths. A No. 8 metric chain is equivalent to an ASI 40.

Alloy steel chains are surprisingly strong. A No. 40 bicycle chain can withstand a ton and a half of force before it elongates and snaps. But for reasonable life, force should be limited to 800 lb or less.

Chains should be cleaned periodically and by soaking in solvent. Most small power transmission chains employ a master link with a removable plate secured by a spring clip. The chain may need to be flexed slightly to free the side plate.

Like cold remedies, there are many chain lubricants, each claiming superiority. Molybdenum disulfide is an old standby, available in aerosol form where it is mixed with a light oil carrier. Apply MoS_2 to the rollers, and when it dries, brush a thick coat of motor oil on the chain to inhibit rust. Castor oil, the original engine lubricant, also makes an effective rust inhibitor. But the type of lubricant is less important than frequent application.

Too much chain tension wears the chain, sprocket, and bearings. Too little tension results in flutter, snatch, and slip (not to be confused with the name of a law office). A 1/4 in. of free play between shaft centers is correct for the sort of applications discussed here. For vehicles with rear suspension, this measurement should be made with the rider aboard. Because suspension pivots rarely share the same center as the drive sprocket, chain tension varies with load.

Chains wear out early because of sprocket misalignment, insufficient tension, and inattention to lubrication. Chains elongate with wear, but do not stretch like taffy. Instead, the pivoting parts—pins, bushings and the rollers that ride on them—develop play, which is cumulative. A chain that has elongated 2% or more should be replaced in order to protect the sprockets from damage. Chains for light motorcycles and motorized bicycles average about 100 1/2-in. links. Moving the rear wheel back 1 in. represents the wear limit.

Sprockets

For most applications, the drive sprocket is about a quarter of the diameter of the driven sprocket and, as a result, wears about four times faster. One way to detect sprocket wear is to wrap a new chain over it. There should be some clearance between the chain and teeth, but not enough to be felt when the chain is tugged.

As the chain wears, the pitch increases. Rollers, now spaced too widely for the sprocket, bite down hard on the trailing edges of the teeth. Once past the surface hardening, wear progresses rapidly. The undercut teeth form themselves into hooks that snag on the chain and are quickly ground away.

While there's not much we can do about manufactured equipment, people who design their own gear can extend chain life with a few simple rules. The most basic of these rules is that the small sprocket should have at least 120° of chain wrap. To say this in a slightly different way, nine teeth in contact is the minimum, with 15 teeth preferred.

The chain-wrap requirement affects other aspects of the design. To maintain proper tooth contact, the speed ratio should be no more than 6:1 and the distance between sprocket centers should be between 30 and 50 chain pitches. If you are using 41 chain (as strong as 40, but narrower), the center distance should be no less than 15 in. and no more than 25 in.

Geared drives

Gearboxes for portable tools are simple devices that, for the most part, consist of a pair of spur gears riding on antifriction bearings. The information supplied here should more than suffice. Multispeed gearboxes used on light motorcycles and on some racing go-karts are another matter. For these, you need a factory manual.

Disassembly

When the driven component—trimmer head, sprocket, auger bit—rotates clockwise, its fastener nearly always has a left-hand thread. Otherwise, the fastener would work itself loose.

In order to unscrew the nut, the shaft must be prevented from turning. How this is done varies. Some shafts have flats for wrench purchase, others lock with a punch as shown in (Fig. 7-11). Dolmar trimmers/brush cutters set the standard with a push-button stop (Fig. 7-12). When all else fails, reach for the air wrench.

Remove all traces of dirt and grease from external surfaces and place the unit on a clean bench for disassembly.

Some output shafts are sealed against dust and water entry. The protective seal may be a separate component or it may incorporate an oil seal on its inner side. The seal will be destroyed during removal; so before you begin, record the number, which should be visible. Also note the shape of the elastomer lip: a seal designed to fend off external threats has the steep side of its lip outboard, away from the gearbox. An oil seal will have its steep side facing inward.

Notches on the unit pictured in Fig. 7-13 enable the seal to be jimmied out from around the shaft. If the manufacturer has not been so considerate, collapse the seal with a dull punch. Flatten the metal face so that the rim pulls away from the seal boss. Work slowly, exercising care not to score the boss or the shaft.

Snap rings, usually backed by spacers, locate the shafts, the bearings in their bosses, and set the gear mesh. Remove the rings with the appropriate tool that, for deep-seated snap rings, may take some hunting to find (Fig. 7-14). Lay out the spacers in order of disassembly.

Antifriction bearings are press-fitted into the aluminum case. Gear cases that split along the centerline of the bearing make disassembly easy (Fig. 7-15). But if the case encircles the bearing, the interference fit must be overcome. Loss of interference fit means the bearing has spun. The surest repair is to replace the gear case, but green Loctite 2701 works for moderate oversizes on the order of 0.003 in.

If the shaft is hollow and accessible from below, a bolt and nut can serve as a puller (Fig. 7-16). But the safest, least destructive way to remove bearings is to heat the housing to around 250°F (121°C). For safety, flush the gear case with nonflammable solvent followed by a water rinse. Spray the disassembled parts with WD-40 to prevent rust.

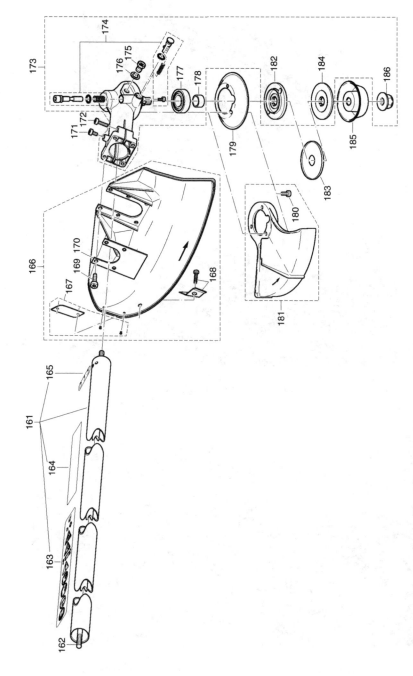

FIG. 7-11. *Dolmar's spring-loaded shaft lock (shown at 174) is a real convenience.*

175

FIG. 7-12. *Most trimmer heads can be locked with a small punch.* Robert Shelby

FIG. 7-13. *Notches on the side of the case simplify seal removal.* Robert Shelby

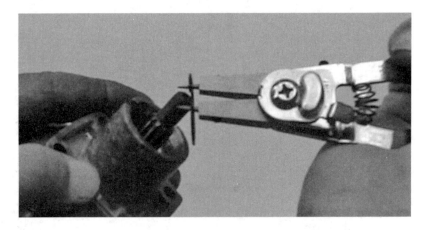

FIG. 7-14. *Snap rings and shims locate shaft assemblies.* Robert Shelby

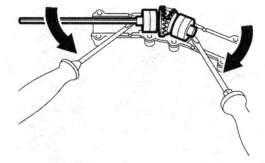

FIG. 7-15. *STIHL angle-drive comes apart easily. A special gear lubricant for this application is available from factory dealers.*

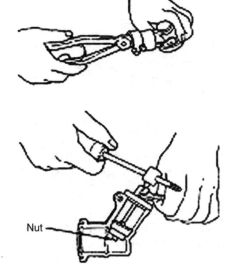

Nut —

FIG. 7-16. *Bearing pullers can be cobbled together from nuts, washers and allthread. Vise-Grips, fitted with a slide hammer that replaces the adjustment screw, comes in handly.* Hitachi Koki USA

Unless the bearings will be replaced, leave them attached to the shafts (Fig. 7-17). Worn bearings can be driven off with a mallet (Fig. 7-18) or freed with moderate heat.

Assembly

A light bulb gives off enough heat to expand bearing inner races for a slip fit with their shafts. Note that sealed bearings go on with their closed faces out. Heat the gearbox as before to insert shafts and bearings. As the second shaft is installed, rotate it as necessary to bring the gear teeth into mesh.

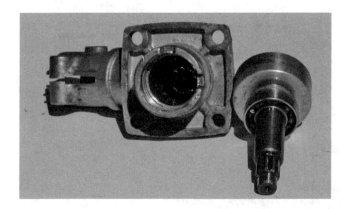

FIG. 7-17. *Unless they are to be replaced, leave the bearings attached to their shafts.*
Robert Shelby

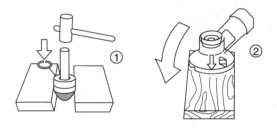

FIG. 7-18. *Old bearings can be driven off their shafts with a soft mallet. If the bearing remains in the housing, apply heat and rap the housing against a wood block.*
Hitachi KoKi USA

Fill the gearbox with the manufacturer's recommended oil. Otherwise, use Mobil-1 75W-90 or an equivalent synthetic gear oil.

Friction drive

I could not sign off without mentioning the Velosolex moped. Home builders can learn a great deal from this machine that, more than any automobile, motorized France. Like the Model T, the Velosolex is a study in simplicity (Fig. 7-19).

The engine mounted over the front wheel and drove through a friction roller. This arrangement gave the necessary gear reduction in one swoop, without recourse to a jackshaft. A handlebar lever raised and lowered the engine into contact with the wheel, although a centrifugal clutch was included in the package.

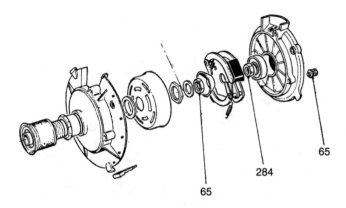

65

284

65

FIG. 7-19. *Velosolex drive in exploded view. Numbers are torque specifications in in/lb. The engine mounted over the front wheel and turned with it, a configuration that gave a pendulum-like persistence to the steering. On the other hand, the mechanicals were well clear of the rider and easy to adapt to bicycle frames. Homebuilders might reconsider front-wheel friction drive, now that engines weigh half as much as the old Velo.*

Because it's less than positive, friction drive reduces wear on the clutch, which is a perennial problem with homebuilt motorbikes. An automotive suspension bushing makes a suitable roller, although some builders prefer to use stacked disks cut from leather or old tires.

Sign off

We've pretty well covered two-stroke engines, how they work and what to do when something goes wrong. I hope readers find the information useful and in the best sense of the word empowering. And now it's time to get back to the shop.

Index